Samuel Hughes, Jas. E. Dennis

A Primer of Map Geography

for Pupils Preparing for Promotion Examinations

Samuel Hughes, Jas. E. Dennis

A Primer of Map Geography
for Pupils Preparing for Promotion Examinations

ISBN/EAN: 9783743334984

Manufactured in Europe, USA, Canada, Australia, Japa

Cover: Foto ©berggeist007 / pixelio.de

Manufactured and distributed by brebook publishing software
(www.brebook.com)

Samuel Hughes, Jas. E. Dennis

A Primer of Map Geography

W. J. Gage & Company's Examination Primer Series.

A PRIMER

OF

MAP GEOGRAPHY

FOR

Pupils preparing for Promotion Examinations.
Pupils *preparing for Entrance Examinations.*
Pupils preparing for Intermediate Examinations.
Students preparing for Teachers' Certificates.
And all Official Examinations.

WITH RECENT DEPARTMENTAL EXAMINATION PAPERS FROM THE PROVINCES OF
ONTARIO, MANITOBA, AND NOVA SCOTIA.

COMPILED BY

SAMUEL HUGHES,
First English Master, Toronto Collegiate Institute,

AND

JAS. E. DENNIS,
Head Master, Woodstock Model School.

W. J. GAGE AND COMPANY
Toronto and Winnipeg.

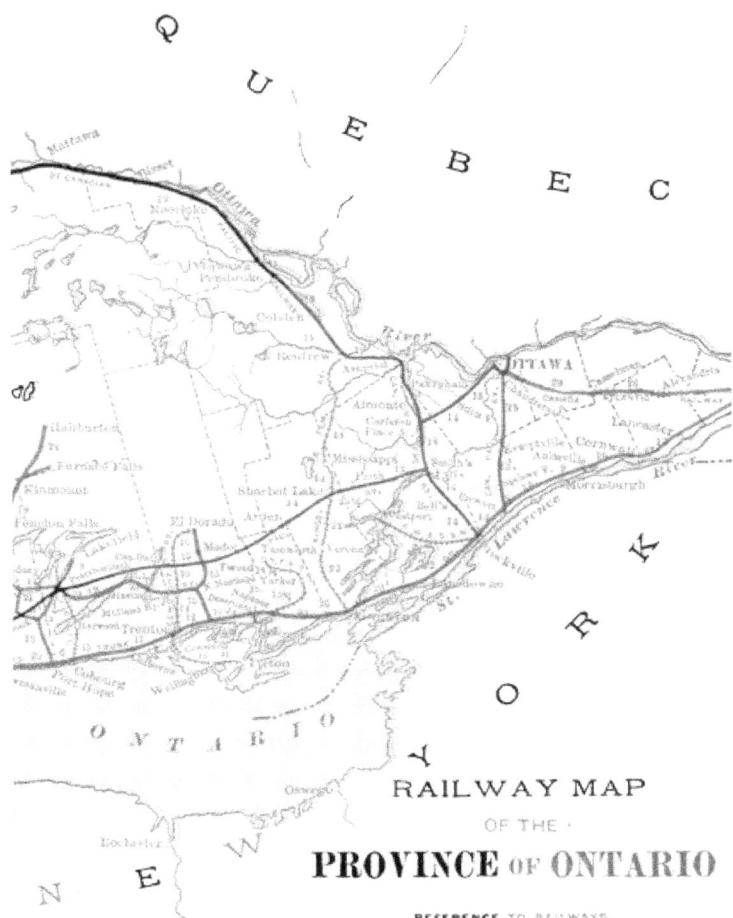

QUEBEC

OTTAWA

RAILWAY MAP

OF THE

PROVINCE OF ONTARIO

NEW YORK

ONTARIO

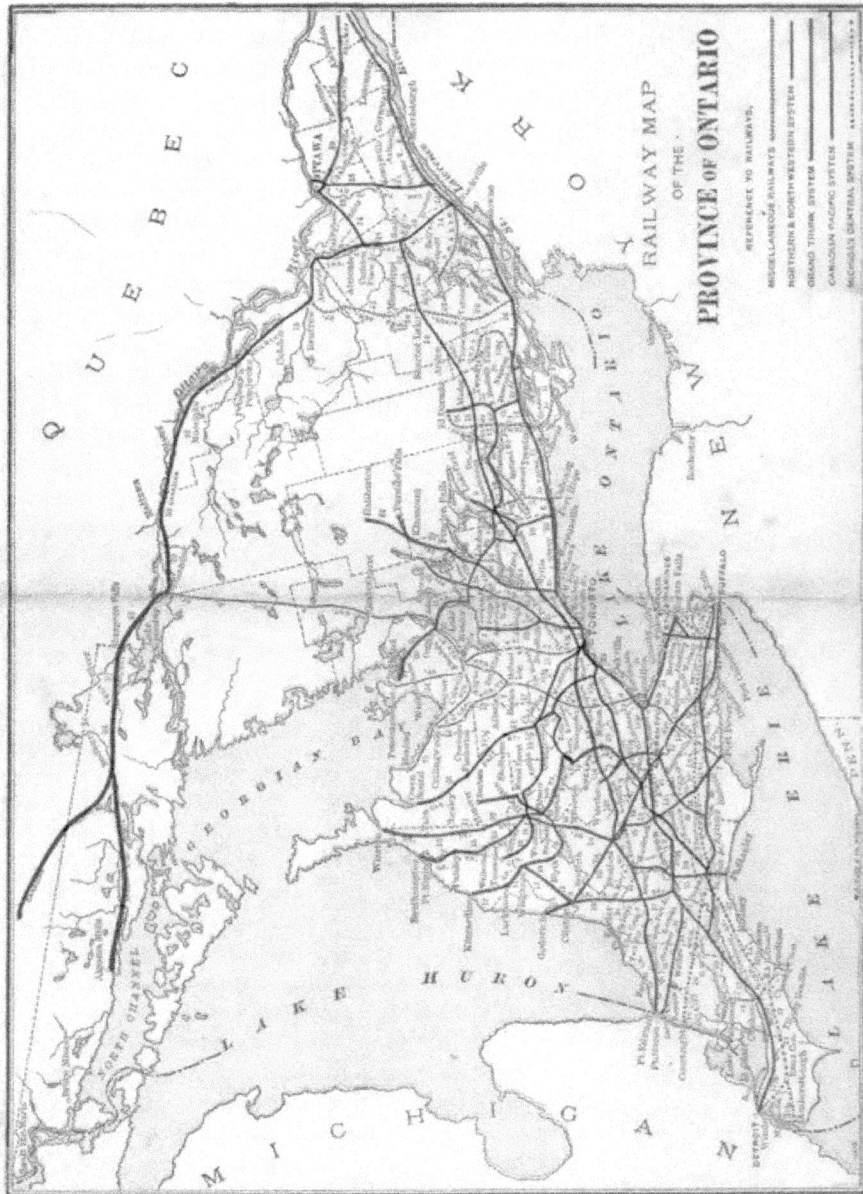

RAILWAY MAP
OF THE
PROVINCE of ONTARIO

REFERENCE TO RAILWAYS.

MISCELLANEOUS RAILWAYS
NORTHERN & NORTH-WESTERN SYSTEM
GRAND TRUNK SYSTEM
CANADIAN PACIFIC SYSTEM
MICHIGAN CENTRAL SYSTEM

QUEBEC

NEW YORK

LAKE ONTARIO

LAKE ERIE

GEORGIAN BAY

LAKE HURON

MICHIGAN

PREFACE.

In compiling this Primer of Geography the aim of the editors has been, not to produce a literary work, but to present, in a simple and systematic form, all the material necessary for the various *promotion, entrance, intermediate,* and *other official* examinations.

To overcome the great difficulty of preparing students for these examinations, some masters throughout the province have taken the ordinary text-books in use, and from the multitude of sentences, selected what they deemed necessary to be learned by the pupil. Others again have used the blackboard or the dictation book for the facts to be memorized. These plans are objectionable, as the one does not present the words from the text-book so as to be remembered readily, and the others necessitate the loss of much valuable time. Further, the pupil does not recognize the word in its written form, and thus the spelling is not taught. Of still more importance, and what of itself should commend this work to teachers and the public generally, is that *the exercise book required for the dictation exercises* in Geography alone *costs as much as this Primer.*

The work is arranged in tabular analyses, to prevent the waste of time in poring over a prosy text-book. Brief notes are inserted at intervals to convey information of special interest. Although merely preliminary, this book will be found to contain all that is necessary to fit a student for any of our examinations in the subject, Geography.

As to *what* and *how much* to teach, those in charge must exercise their own judgments.

Since many geographical terms are spelled two ways, it has been the endeavor to present the more general form, for instance: Listowell, Chippawa, Deseronto, Porto Rico, Mines, Guadalquivir, &c., &c.

The attention of both teacher and student is directed to the Railway Map and to its analysis as special features of the book.

The thanks of the editors are due to the Hon. Mackenzie Bowell, Minister of Customs, Ottawa, for official statistics, reports, etc.; to Robert Jaffray, Esq., Director of the Midland Railway system, Toronto, to J. H. Knight, Esq., Inspector of Schools, Lindsay, and to others for maps and facts in connection with the railways.

INDEX.

GEOGRAPHY.

Geography.	A description of the earth. There are three departments : **Mathematical, Physical, Political**.
Mathematical.	The form, motions, and magnitude of the earth.
Physical.	The natural divisions of the earth's surface into land and water, the wind, rain, atmosphere, etc., etc.
Political.	The artificial divisions, i.e., into countries, cities, governments, **etc.**

Mathematical Geography Definitions.

Earth.	A planet, i.e., a cool body revolving around an incandescent one.
Shape of Earth.	An oblate spheroid ; i.e., a slightly compressed sphere.
Motions.	Diurnal, or Daily, i.e., turning on its axis from W. to E. once in 24 hours (23 h., 56 m., 4 sec.). Result, day and night. Annual, or Yearly, i.e., revolving around the sun in 365 dys., 6 hrs., 9 min., 10·7 sec. Result, Seasons : Spring, 93 days ; Summer, 94 days ; Autumn, 89 days ; Winter, 89 days. Rate of travel, 17,000 miles an hour. Universal, moving through space as part of the solar system.
Gravity.	The reciprocal attraction of matter to matter. Note—The centre of gravity is south of the centre of the earth. Result, more water in the southern than in the northern hemisphere.
Axis.	The line about which the earth turns. Note—All lines referred to in these definitions are imaginary.
Diameter.	A straight line passing through the earth's centre and terminating at both ends at the surface. Equatorial diameter, 7,925·6 miles ; Polar, 7,899·2 miles. In general terms, 8,000 miles ; Circumference, 24,856, or say 25,000 ; Area, 197,000,000 miles ; Solid contents, 260 billions of miles.
Poles.	The ends of the axis. There are two : North and South.
Cardinal Points	East, where the sun rises ; West, where it sets ; South, towards the horizon directly under the sun at noon ; North, opposite the South, or towards the polar star.
Horizon.	Sensible, where the sky and earth seem to meet. Rational, a great circle of the celestial sphere whose plane passes through the centre of the earth, and whose poles are the zenith and the nadir. It is parallel to the sensible horizon. Zenith, the point directly overhead ; nadir, the point directly underneath.
Equator.	A great circle around the earth, midway between the poles. This divides the earth into Northern and Southern hemispheres, and is the base line whence latitude is calculated.
Latitude.	Distance north or south of the equator. Degrees of latitude are all very nearly the same length.
Meridian.	A semi-circle passing from pole to pole and cutting the equator at right angles.
Longitude	Distance east or west of the first or base meridian. The meridian of Greenwich, Eng., is the base. All east of it for 180° is E. Longitude ; all west of it for 180° is W. Longitude. 360° makes a difference of 24 hours, or 1° makes 4 min. Degrees of Longitude vary in length from 69¼ miles to 0.
Parallels of Latitude.	Small circles parallel to the equator.
Important Parallels.	Tropic of Cancer 23° 28′ N.; Tropic of Capricorn, 23° 28′ S.; Arctic, or N. Polar Circle, 23° 28′ from the N. pole ; Antarctic, or S. Polar Circle, 23° 28′ from the S. pole.
Great Circle.	One dividing the earth into two equal parts, i.e., one whose plane passes through the earth's centre. They are : equator, ecliptic, meridian, rational horizon, etc. Small circles do not divide the earth equally.
Zones, or Belts.	Torrid, between 23° 28′ S. and 23° 28′ N., i.e., within the Tropics. Temperates, between the Tropics and the Polar circles, each 43° 4′ wide. Frigids, within the Polar circles, each 46° 56′ in diameter.
Ecliptic.	The great circle described by the earth in a year. Its plane cuts the equator at an angle of 23° 28′. Eclipses occur only when the moon is on or near this circle.
Circle of Illumination.	The line dividing light from darkness.
Eclipse.	Sun : moon comes between earth and sun. Moon : earth comes between moon and sun.

Equinoxes.	When the sun is opposite the equator, i.e., occupies the intersecting points of the ecliptic and the equator. **Vernal Equinox**, 21st Mar.; Autumnal, 23rd Sept. Day and night are equal all over the world at these periods.
Solstices.	When the sun is most remote from the equator, i.e., when it reaches the tropics and seems to stand before returning. **Summer Solstice**, 21st June; Winter, 21st December.
Zodiac.	A belt in the heavens 9° on each side of the ecliptic. The chief planets revolve in it.
Moon.	A secondary planet, or satellite, i.e., a small body revolving around a planet. Our moon reflects the light of the sun, and is the chief cause of tides. It completes one revolution in 27 dys., 7 hrs., 45 m., but new moon occurs only in 29 days, 12 hrs. 44 min., a difference of 2 dys., 5 hrs.; *Cause, the earth's moving on in the ecliptic.* A place on the earth directly under the moon at a certain time is not in the same relative position 24 hours afterwards, but is more than 12° short, i.e., the distance traversed by the moon in its orbit. Therefore the earth must turn on 51 min. to bring the place in the same relative position. Hence on the average the moon rises 51 min. later each succeeding day, or one day in each lunar month. The tides are also 51 min. later, i.e., there are two tides in 24 hrs., 51 min. *Moon's mean distance, 237,000 miles; diameter, 2,153 m.; size, 50 times smaller than the earth; density, but little more than half the earth's, hence its attractive force is only 1/50 as great.*
Sun.	The source of light and heat. The centre of our universe. *Distance, 91,000,000 miles; size, 1,300,000 times larger than the earth.*
Declination.	Distance of the sun north or south of the equator.
Altitude.	Distance above the horizon.
Map.	A plan of the earth or of a part of it. The top of the map is North, the bottom South, the right side East, the left West.

Physical Geography Definitions.

Land covers 52,000,000 miles, or about ⅓ of the earth's surface, chiefly north of the equator.

Continent.	A large mass of land. There are six:—Europe, Asia, Africa, N. America, S. America, Oceania. Europe, Asia, and Africa are termed the Old World; North and South America the New World.
Island.	Land surrounded by water. **Islet, a small island**; Group or Archipelago, several islands close together.
Peninsula.	Land almost surrounded by water.
Isthmus.	A strip of land joining two large bodies of land.
Coast, or Shore.	Land bordering on the water. Particular names:—*Seaboard, strand, beach, cliff, bluff, bank.*
Cape.	Land jutting into the water. Local names:—*Head, point, ness, naze, mull, bill, promontory, butt.*
Hill.	An elevation of less than 2,000 feet above the surrounding country. Other names:— *Hillock, knoll, dune, down, mound, tor, cap, beacon, low.*
Mountain.	An elevation of more than 2,000 feet above the adjoining country. Summit, top, or peak is the highest part; foot or base, the lowest; precipice or escarpment, a steep slope. Local terms:—*Ben, pen, berg, mont, alps, sierras, cordilleras, andes.* Range or chain, a continuous line of mountains; pass, defile, or cañon, a narrow opening in a range; glacier, a huge mass of ice on a mountain; moraine, a mass of earth loose on a mountain; avalanche, a snow slide from a mountain; volcano, a burning mountain; crater, the opening in a volcano; earthquake, a shaking of the earth's crust.
Plateau, or Tableland.	An extent of land more than 1,000 feet high, nearly uniformly elevated above the sea.
Highland.	A series of irregular elevations with valleys between.
Plain.	Level country less than 1,000 feet high. Local names:—*Landes (sand heaths), steppes (barrens), deserts (sandy), prairies (grassy), savannas (wet), silvas (wooded), llanos (grassy and wood), pampas (treeless), moor (heath).*
Valley.	A depression between hills and mountains. Particular names:—*Glen, ravine, gorge, strath, dale, vale, cañon, gully.*
Delta.	Land between the mouths of a river. **Bar**, a bank across the mouth of a river.
	Water covers 145,000,000 miles, or ⅔ of the earth, chiefly south of the equator.
Ocean.	A very large division of water. These are: Pacific, Atlantic, Indian, Antarctic, and Arctic.

Sea.	A large branch of an ocean. In general terms *sea* means ocean, sea, gulf, lake, etc.
Gulf.	Part of a sea or an ocean extending into the land.
Bay.	A hollow, bend, or indentation in the coast line. Bight is an open bay; Inlet, a general name for all coast openings.
× **Port.**	Inlet affording shelter to ships. Other names : *Harbor, haven.*
Roadstead.	A sheltered place for ships to swing or ride at anchor.
Firth, or Frith.	A narrow inlet at the mouth of a river. Fiord is the Scandinavian name.
Estuary.	The part of a river affected by tides.
Marsh.	Low wet land. Lagoon, a marsh near the sea coast. Swamp, a wooded marsh. Bog, a marsh of vegetable deposit.
Strait.	A passage between bodies of land. Sometimes called Sound.
Channel.	A wide strait ; also, where a river expands into the sea.
Current.	The progressive motion of waters. Wave, billow, surge, swell, are rolling waters caused by the winds, tides, &c.
Tides.	The regular rising and falling of waters in oceans, bays, &c. Causes : the attraction of the sun and moon. Flood is the rising tide ; Ebb, the falling ; Spring tide, at the new and at full moon ; Neap, at the moon's *first* and *third* quarters.
Lake.	A large body of water surrounded by land. There are four classes : (*a*) Those that only *receive* waters; (*b*) those that only *give out* waters; (*c*) those that *both receive and give out*; (*d*) those that *neither receive nor give out.* Local names :—*Loch, lough, mere, tarn, water, sea, lac,* &c.
Spring.	Water coming through the earth's surface.
River	A large fresh water stream, flowing into a sea, lake, &c. Source or head, where the remotest part rises ; Mouth, where the waters reach the sea ; Bed, the channel in which the waters flow ; Banks, the sides of the stream. The *right bank* is on one's *right side* going down stream, the *left bank* is on the *left.* Affluent, tributary, branch, fork, feeder, &c., a stream flowing into a river. Confluent, one entering the sea at the same place as another ; Confluence, where two rivers join ; Basin, the whole area drained ; Watershed, the ridge separating basins ; Rapid, a swift current ; Fall, cataract, cascade, the descent of a stream over a precipice.
Canal.	An artificial river for purposes of navigation, drainage, irrigation, &c.
Ocean Currents	Regular movements of ocean waters. Causes : *Evaporation, wind, rotary motion, differences in specific gravities and in temperatures of waters, tides.*

Political Geography Definitions.

Republic.	Where the *executive* and *legislative* powers are exercised by persons *elected* by the people.
Monarchy.	Where the executive power is vested for life in one person who usually inherits the office. Kinds : Empire, kingdom, principality, &c. *Limited Monarchy* : people's representatives usually control both legislative and executive powers, sovereign acting on the advice of counsellors chosen from the representatives. *Absolute Monarchy* : The sovereign is unrestricted.
Colony.	A settlement in a foreign land by people emigrating from their mother country.
Occupations of the Human Race.	*Agriculture, Stock-raising, Mining, Lumbering, Manufacturing, Commerce, Fishing,* and *Hunting.*
Civilized Nations.	Such as are governed by laws emanating from the people.

Note.—In the text R stands for *republic* ; K, *kingdom* ; E, *empire* ; P, *principality.*

British Colonies.

Note.—Those in black type have *responsible* government; those in *italics, representative* ; those in Roman, *crown.*

European.	(Cyprus), Gibraltar, Heligoland, Malta, Channel Islands, Isle of Man.
American.	**Canada.** Newfoundland, *Bahamas, Bermudas,* Honduras, Jamaica and Turks, *Leewards, Windwards,* Falklands, Guiana, Trinidad.
African.	Ascension, Cape of Good Hope and dependencies, Gambia, Gold Coast, Lagos, Mauritius, *Natal,* St. Helena, Sierra Leone.
Asiatic.	Aden, *Ceylon,* Cyprus, Hong Kong, India, Labuan, Perim, Straits Settlements.
Australasiatic.	Fiji Islands, Rotumah, **New South Wales, New Zealand, Queensland, South Australia, Tasmania, Victoria,** *Western Australia.*

NORTH AMERICA

Area, 8½ millions of miles; length, 5600 m.; breadth, 3120 m.
Coast line, 28000 m.; Latitude, 10°-80° N.; Longitude, 55°-165° W.
Mean height above the Ocean, 700 feet,

BOUNDARIES.
{
North:—Arctic Ocean and Baffin Bay,
West:—Pacific Ocean and Behring Sea,
East:—Atlantic Ocean.
South:—Caribbean Sea and Pacific Ocean
}

Country.	Comp. Size.	Area 1000s of sq. mls.	Gov't.	Capital and its Location.	Exports
United States	26	3160	R	Washington on the Potomac	Breadstuffs, raw cotton, and manufactured goods of cotton and of iron, etc.
ALASKA	5	600		Sitka on Sitka island	Fish, seals, furs and fish-oils.
Canada	28	3370	Resp	Ottawa on the Ottawa	Lumber, timber, breadstuffs, manufactural goods.
Newfoundland (including Labrador)	1-3	40	''	St. John's on St. John's bay	Fish, oils, seals, furs, etc.
Jamaica and Turks		6	Cr'wn	Kingston on Port Royal	Sugar, rum, tobacco, fruit.
Brit. Honduras, or BELIZE	1-9	13	''	Belize on Honduras bay	Fruit, coffee, mahogany and other cabinet woods.
Bermudas		1-42	Repr	Hamilton on Long Island	Potatoes, bananas, oranges
Bahamas		5½	''	Nassau on New Providence	Salt, sponges, oranges, pineapples
Brit. Leewards		1	''	St. John on Antigua	Molasses, rum, sugar, arrowroot, cotton, and tobacco.
" Windw'ds		1	''	Bridgetown on Barbadoes	Sugar, molasses, rum, and turtles.
Guatemala	1-3	40	R	New Guatemala on the Montagua	Coffee, cochineal, mahogany, sarsaparilla and dyewoods
San Salvador		7	''	San Salvador on the coast	Indigo, coffee, tobacco, sugar, balsam, hides, rice, cedar.
Honduras	1-3	40	''	Tegueigalpa on the Choluteca	Fruits, cotton, sugar, tobacco, indigo, rosewood.
Nicaragua	1-3	58	''	Managua on lake Leon, or Managua	Fruits, cotton, sugar and tobacco.
Costa Rica	2-0	20	''	San Jose on the Carthago	Sugar, tobacco, corn, cocoa, dyewoods and fruits.
San Domingo, or DOMINICA	1-6	20	''	San Domingo on the south coast	Sugar, cotton, lime-juice, cabinet woods.
Hayti and Tortuga	1-12	10	''	Port-au-Prince on the west coast	Mahogany, logwood, honey, coffee, cocoa.
Mexico	6	742	''	Mexico near lake Tezcuco	Fruits, dye- and cabinetwoods, medicines, indiarubber, gold, silver, tobacco, coffee, hides.
Danish Colonies:					
GREENLAND	3-8	46	Cr'wn	Godshaab, Uppernavik, etc.	Eider-down, seal skins, whalebone, and oils.
ICELAND	1-3	44	Repr	Reikiavik	
ST. CROIX } ST. THOMAS } ST. JOHN }			Cr'wn	St. Thomas and Christianstadt	Unrefined sugar, raw cotton.
Spanish Colonies					
CUBA	1-5	43	Repr	Havana on Cuba	Raw sugar, tobacco, cigars,
PORTO RICO		3		San Juan on Porto Rico	
French Colonies:					
MIQUELON } ST. PIERRE }		1-5	''	St. Pierre	Fish.
GUADALOUPE		2	''	Basse Terre	Sugar, coffee, cocoa, fruits.
MARTINIQUE		1	''	Port Royal	
Dutch Colonies :		1-3	Cr'wn	Williamstadt on Curaçoa	Fine woods, dyes, fruits, cattle, salt.

Note 1.—The British Isles, area 121,000 miles, is the basis of comparison.

2.—**Crown Colony**: The CROWN, or Home government, has the entire control both of *legislation* and of *administration*.

Representative government colony: The CROWN retains merely a veto on *legislation*, but controls the *administration*; *i. e.*, appoints all officers to execute the laws.

Responsible government colony: The CROWN retains merely a veto on legislation, and has no control over any executive or administrative officer except the governor.

3.—The imports of the above countries are chiefly manufactured cottons, woollens, hardware, &c.

4.—Iceland is by some considered a part of Europe.

Straits Sounds, and **Channels.**	*On the north :*—Smith, Jones, Lancaster, Barrow, Melville, Banks, Victoria, Franklin, McClintock, Fury and Hecla, Dease, and Fox. *On the east :*—Davis, Hudson, Frobisher, Belle Isle, Northumberland, Canso, Long, Florida or Bahama, Yucatan, Windward, and Mona. *On the west :*—Juan de Fuca, Haro, Rosario, Georgia, Johnston, Broughton, Queen Charlotte, Scott, and Behring.
Gulfs and Bays.	*Arctic :*—Baffin, Disco, Melville, Boothia, Coronation. *Atlantic :*—Hudson, James, Ungava, Penny, St. Lawrence, Chaleur, Chedabucto, Halifax, Fundy, Passamaquoddy, Mines, Annapolis, Chignecto, Massachusetts, Cape Cod, Delaware, Chesapeake, Charleston, Appalachee, Mexico, Campeachy, and Honduras. *Pacific :*—California, San Francisco, and Georgia.
Islands.	*Off the north coast :*—Greenland, the *Arctic archipelago*, and Cumberland. *Off the east coast :*—Iceland, Cockburn, Fox, Southampton, Newfoundland, Anticosti, Magdalens, Prince Edward, Cape Breton, Miquelon, St. Pierre, Sable, Nantucket, Martha's Vineyard, Long, Bermudas, Bahamas, Keys (Caicos, or Turks), Cuba, Dominica (Hayti and San Domingo), Jamaica, Porto Rico, Virgins, Windwards, and Leewards. *Off the west coast :*—Vancouver, San Juan *archipelago*, Scotts, Queen Charlotte, Prince of Wales, Sitka, Aleutians, Kodiac, St. Lawrence, and Santa Barbaras.
Peninsulas.	*North :*—Boothia and Melville. *East :*—Labrador, Nova Scotia, Gaspé, Cape Cod, Maryland, Florida, and Yucatan. *West :*—California and Alaska. NOTE.—Nearly all peninsulas, except Yucatan, Labrador, and Jutland (Denmark), point southerly. *Cause :*—The *centre of gravity* is south of the centre of the earth ; hence more of the southern hemisphere is submerged.
Capes.	*On the Arctic seaboard :*—Lisburn, Icy, Barrow, Demarcation, Bathurst, and Murchison. *On the Atlantic seaboard :*—Farewell, Chudleigh, Wolstenholm, Henrietta Maria, Jones, Charles, Ray, Race, Gaspé, Canso, Breton, Sable, Ann, Cod, May, Henlopen, Charles, Henry, Hatteras, Tancha (*Sable*), Catoche, and Gracias a Dios. *On the Pacific seaboard :*—Corrientes, San Lucas, Conception, Mendocino, Blanco, Flattery, Cook, Scott, Romanzoff, and Prince of Wales.

MOUNTAINS.

Range.	Location.	Highest Peak.	Height in feet.	Latitude.
Rocky	From the Arctic Ocean to the Isthmus of Darien, along the western-central part of the continent.	Brown / Hooker / Popocatapetel	15,900 / 15,000 / 17,883	52° 35 / 52° 15 / 19° 53
Coast	From the mouth of the Fraser river to Alaska, along the coast.	St. Elias	17,900	60° 20
Cascade	From the Fraser river mouth to southern California, along the coast.	Fairweather / Hood	14,750 / 17,000	59° 2 / 45° 4
Nevada	Between California and Nevada.	Shasta / Whitney	14,400 / 15,000	41° 6 / 36° 32
Humboldt	In Nevada.	....	14,800	......
Sonora	In Mexico, along the Gulf of California.	Colima	12,000	......
Ozark	In Arkansas and Missouri.	....	3,000	......
Appalachian, or Alleghany	The main ridge in Tennessee and West Va. The main ridge in Pennsylvania.	....		
Iron	In western North Carolina.	Unaka	6,000	
Cumberland	In eastern Tennessee and Kentucky.	Look Out	5,000	
Blue	In West Virginia, Maryland, & Pennsylvania.	Mitchell / Otter	6,400 / 6,500	
Green	In Vermont.	Mansfield	4,430	44° 35
White	In New Hampshire.	Washington	6,400	44° 25
Catskill	In New York, west of the Hudson.	....	3,800	
Adirondacs	"　　" 　　" 　lake Champlain.	Marcy	5,400	
Lawrence, or Laurentian	"　　" 　southeast of the river St. Lawrence.	....	3,000	

Note. 1.—The West Indies are all mountainous, many peaks being from 6,000 to 7,000 feet high. Hecla, in Iceland, is 5,100 feet in height.

Note. 2.—The *Rocky* system is 5,000 miles long and from 570 to 1,040 wide ; average height 5,000 feet. The *Alleghany* system is 2,000 miles long and from 150 to 200 wide ; average height 2,500 feet.

Plains.　　1. *Mississippi* valley, including the area drained by the Missouri, the Ohio, and other branches of the Mississippi.

2. *Saskatchewan,* comprising all the country drained into lake Winnipeg.

3. *Mackenzie,* or *Great Northern,* sloping towards the Arctic Ocean.

4. *Atlantic,* or *Great Eastern,* embracing all east of the Alleghanies.

PLATEAUX.

Name.	Remarks.	Name.	Remarks.
1. Labrador	Average height 2,000 feet.	8. Nevada	Average height 8,000 feet.
2. Laurentian	In northeast Ontario and Quebec.	9. Gt. Western	In Nebraska, Colorado, and New Mexico.
3. Acadian	In New Brunswick.	10. Arizona	Average height 5,500 feet.
4. C. bequid and North	In Nova Scotia.	11. H'ght of Land	Along 49th parallel.
5. Ontario	Between lake Huron and the Ottawa.	12. Anhuac or Mexican	Average height 7,000 feet.
6. Maine	Average height 5,000 feet.	13. Guatemala	"　　" 　5,000　"
7. Utah	"　　" 　8,000　"	14. Gt. Eastern	Among the Alleghanies.

Rivers.

Into the Arctic Ocean :—Colville, Mackenzie, Coppermine, Great Fish (Back)

Into Hudson Bay :—Churchill (Great English), Nelson, Hayes, Severn, **Whale.**

Into James Bay :—Albany, Moose, Abittibi, Rupert, East Main.

Into **St. Lawrence** *Gulf* : -St. Lawrence, Restigouche, Miramichi, Richibucto.

Into **Bay of Fundy :—Petitcodiac,** Kennebeccasis, St. John, St. Croix.

On the Atlantic Slope } Penobscot, Kennebec, Androscoggin, Merrimac, Connecticut,
of the United States : } Hudson, Delaware, Susquehanna, Potomac, Rappahannock,
James, Roanoke, Cape Fear, Pedee, Santee, Savannah, Altamaha, St. John's.

Into the Gulf of Mexico :—Appalachicola, **Alabama,** Mobile, **Mississippi, Sabine,**
Trinity, Brazos, Colorado of Texas, Grande del **Norte.**

On the Pacific Slope :—Colorado of California, San Joaquin, Sacramento, Kalmath,
Columbia, Fraser, **Skeena, Stickeen,** Yukon.

Note 1.—The Mackenzie receives :—

The *Great Bear* from Great Bear Lake, and the *Liard,* **or** *Mountain,* **from Br. Columbia ;**
and also drains Great Slave Lake.

The *Slave* flows **from Lake Athabasca to Great** Slave Lake.

The *Peace, Athabasca,* and *Stone* rivers flow into Lake Athabasca. In Br. Columbia, near
the source of the Fraser, the *Peace* receives the *Findlay* and the *Parsnip,* and in
Athabasca territory the *Smoky* and the *Rouge.* Both the *Columbia* and the *Athabasca*
have their source in "The Committee's Punch Bowl," a lakelet on Mt. Brown. The
Athabasca flows northerly through Alberta and Athabasca territories, receiving the
Lesser Slave from Lesser Slave Lake, and the *Great Pembina.*

Lakes Wollaston and Deer are drained both into Lake Athabasca by the *Stone,* and into
the *Churchill* by the *Great Deer.*

*Total length of the Mackenzie, via the Athabasca, 2300 miles. Area drained, 600000 miles
or 1/5 of Canada.*

2.—The **Nelson** drains Lake Winnipeg. This lake receives the following rivers :—

(a) *Saskatchewan,* with its tributaries: the *North* and *South* branches and their **affluents :** *Bow,
Belly, Red Deer,* and *Battle ;*

(b) *Little Saskatchewan,* from Lakes Manitoba and Winnipegosis ;

(c) *Red of the North,* and its tributaries: *Roseau, Pembina,* and *Assiniboine,* with the affluents
of the latter : *Souris* (Mouse), *Qu'appelle* (Who Calls).

(d) *Winnipeg,* from Lake of the Woods, with its affluent the *English* from Lonely lake (Seul).
Lake of the Woods receives the *Rainy* from Rainy Lake, into which flow the *Seine* and
Namenkan.

*Total length of the Nelson and Saskatchewan, 1700 miles. Area drained, over 360000
miles, or 1/9 of Canada.*

3.—The **St. Lawrence** drains the great lakes. (See "Lakes.")

*Total length of St. Lawrence, via the Great Lakes, 2150 miles. Area drained, over
300000 miles, or 1·11 of Canada.*

4.—The **Mississippi** drains the valley between the Rockies and **the** Alleghanies, from the Canada
border to the gulf of Mexico. Its tributaries are :

From the West :—Minnesota, Des Moines, Missouri, Arkansas, and *Red of the South.*

From the East :—Wisconsin, Illinois, and *Ohio.*

The *Missouri* receives the *Yellowstone, La Platte,* and *Kansas.*

The *Ohio,* formed by the *Allegheny* and *Monongahela* at Pittsburg, Pa., receives :—*Kanawha,
Licking, Kentucky, Cumberland,* and *Tennessee* from the South ; and *Scioto, Miami,*
and *Wabash* from the North.

*Total length of the Mississippi, via the Missouri, 4506 miles. Area drained, 1226600
miles, or 1/3 of the United States.*

5.—The **Columbia** drains part of Br. Columbia, **Washington,** and Oregon. Its tributaries **are :—**
Kootenay, Okanagan, Lewis (Snake), *Des Chutes,* and *Willamette.*

*Total length of the Columbia from "The Committee's Punch Bowl" to its mouth, 1200
miles. Area drained, over 300000 miles.*

Superior

LAKES.

Superior

Name.	Area in Miles.	Length.	Breadth.	Height.	Affluents.	Outlets.
Superior	32000	420	160	627	Kaministiquia, Nipigon, Pic, Michipicoten.	Canal (on United States shore) and river *St. Mary* (Sault Ste. Marie) into lake Huron.
Michigan	22000	360	80	578	Fox, Manistee, Grand, St. Joseph, Kalamazoo.	*Chicago* canal into Illinois river, and straits of *Mackinac* into lake Huron.
Huron, with Saginaw and Georgian bays	24000	240	180	578	Spanish, French, Magane-tewan, Muskoka, Severn, Saugeen, Maitland, St. Mary, Mackinaw.	*St. Clair* river into lake St. Clair.
St. Clair	360	36	24	570	Thames, Sydenham or Bear Creek, St. Clair.	*Detroit* into lake Erie.
Erie	12000	250	80	565	Detroit, Grand, Maumee, Otter, Kettle.	*Niagara* river and *Welland* canal into lake Ontario, and *Erie* canal into Hudson river, and lake Ontario.
Ontario and Bay of Quinte	7000	180	65	232	Niagara, Credit, Trent, Moira, Genesee, Oswego, Black.	*St. Lawrence* river to the Gulf, *Rideau* canal to Ottawa river, and *Genesee* and *Oswego* canals to Erie canal.
Nipigon	1650	60	50	850	....	*Nipigon* into lake Superior.
Simcoe	....	30	18	704	Holland, Beaver, Talbot.	*Severn* into Georgian bay, through lake Couchiching.
Nipissing	....	50	35	634	Sturgeon.	*French* into Georgian bay.
Temiscamingue	....	67	15	656	Ottawa.	*Ottawa* into St. Lawrence.
Champlain	767	104	15	90	Missisquoi, Winooski, and Horicon from lake George.	*Richelieu* (Sorel) into St. Lawrence.
St. John	500	....	....	300	Mistassini	*Saguenay* into St. Lawrence.
Great Bear	14000	200	....	230	....	*Great Bear* into Mackenzie.
Great Slave	20000	350	....	580	Great Slave, Hay.	*Mackenzie* into Arctic Ocean.
Athabasca	5000	240	....	600	Peace, Athabasca, Stone, Deer.	*Great Slave* into Great Slave lake.
Wollaston	2000	....	....	....	Deer.	*Stone* into lake Athabasca.
Deer	3000	115	....	....	Deer.	*Great Deer* into Churchill, and *Deer* into lake Wollaston.
Winnipeg	18000	280	....	700	Saskatchewan, Little Saskatchewan, Red, Winnipeg.	*Nelson* into Hudson bay.
Winnipegosis	2000	120	....	728	Red Deer, Swan, Shoal.	*Water Hen* into Lake Manitoba.
Manitoba	....	....	....	720	White Mud, Water Hen.	*Little Saskatchewan* into lake Winnipeg.
Woods	500	....	....	977	Rainy.	*Winnipeg* into lake Winnipeg.
Lonely, or Seul	....	....	....	....	....	*English* into Winnipeg river.

Minor lakes:—Tezcuco and Chapala, in Mexico; **Nicaragua** and Leon (*Managua*), in Nicaragua; Great Salt and Utah, in Utah; Humboldt, in Nevada; Itasca, in Minnesota; Okanagan, in British Columbia; Abittibi, Balsam, Sturgeon, Scugog, Pigeon, Chemong, Stoney, Rice, and Mississippi, in Ontario; **George or** Horicon, in New York; and **Memphramagog**, **Megantic**, **Temiscouata**, and **Manouan**, in Quebec.

Note 1.—The lake expansions of the St. Lawrence are: **Thousand Islands**, St. Francis, St. Louis, and St. Peter.

2.—Lake Superior is 1200 feet deep; Michigan, 1000; Huron, 600; St. Clair, 20; Erie, 200; Ontario, 600.

Canada

Capital, *Ottawa*.   *Population, 4,324,108 (1881).*

Area, 3,370,290 miles.
Latitude : Same as from the north of Norway to the middle of Spain.
Southern extremity, Point Pelee, in Essex, Ontario, in Lat. 42°, or the same as Boston.

BOUNDARIES.

North :—Arctic Ocean; East:—Atlantic Ocean; West:—Pacific Ocean; South:—United States and the great lakes; or more minutely as follows :—

(a) *From the Gulf of Georgia to lake of the Woods :* the 49th parallel of latitude.

(b) *From lake of the Woods to lake Superior :* Rainy river and lake, with a chain of tributary lakes, and Pigeon river.

(c) *From the mouth of Pigeon river to St. Regis, Quebec, opposite Cornwall :* Lakes Superior, Huron, St. Clair, Erie, Ontario, Thousand Islands ; and the rivers St. Mary, St. Clair, Detroit, Niagara, St. Lawrence.

(d) *From St. Regis to the N. E. corner of Vermont (the source of the Connecticut) :* the 45th parallel.

(e) *Thence to Passamaquoddy Bay :* a ridge of highlands and the St. John and St. Croix rivers.

Government.

1. The *legislative* power is a *Parliament* constituted as follows:—

(a) House of Commons, whose members are *elected* by the people of Canada. *Ontario sends 92 ; Quebec, 65 ; Nova Scotia, 21 ; New Brunswick, 16 ; Manitoba, 5 ; British Columbia, 6 ; and Prince Edward Island, 6.* Total, 211. Any bill may, and all money bills must, originate in this House. It controls the revenue and the expenditure.

(b) **Senate, or** Upper House, whose members are appointed for life by the **Privy** Council of Canada. *There are 24 from Ontario ; 24 from Quebec ; 10 from Nova Scotia ; 10 from New Brunswick ; 3 from Manitoba ; 3 from British Columbia ; and 4 from Prince Edward Island.* Total, 78.

> NOTE.—The senators of the first parliament of Canada (1867) received their commissions from the Privy Council of Great Britain and Ireland.

(c) Governor-General, appointed by the Privy Council of Great Britain and Ireland, and representing the Queen.

2. The *executive* power is a *Privy Council* composed of **the** Governor-General and members summoned by him. In *practice* these are : the *leader* of the dominant party in the House of Commons and *members of both Senate and Commons* recommended by him to the Governor-General for appointment.

The administration **of the laws is** entrusted to judges *appointed* by the Privy Council of Canada, and to **magistrates appointed by the** Executive Councils of the Provinces.

The revenue is derived chiefly from : (a) Customs, a duty levied on goods imported ; (b) Excise, a duty levied on liquors, tobaccos, &c., *manufactured* in Canada ; (c) Post Office ; (d) Public Works, &c.

The Expenditure is chiefly on account of : interest on the public debt, subsidies to Provinces, civil government, justice, legislation, penitentiaries, surveys, **public** works, post office, Indian grants, immigration, militia, &c.

Public debt of Canada, $225,000,000. Exports, $100,000,000. Imports, $105,000,000. Revenue, $30,000,000. Expenditure, $27,600,000.

PROVINCES OF CANADA.

Name	Area in thou. of m.	1881. Popula-tion.	Capital and its Population in thousands.	Location of Capital.	Remaining cities and towns with upwards of 5,000 inhabitants.
Ontario	220	1,923,228	Toronto 86	Lake Ontario	Hamilton 36, Ottawa 27, London 20, Kingston 14, Guelph 10, St. Catharines 10, Brantford 10, Belleville 10, St. Thomas 9, Stratford 8, Chatham 8, Brockville 8, Peterborough 7, Windsor 7, Port Hope 6, Woodstock 6, Galt 5, Lindsay 5, Barrie 5.
Quebec	188	1,359,027	Quebec 62	St. Lawrence river	Montreal 140, Three Rivers 9, Levis 8, Sherbrooke 7, Hull 7, St. Henri 6, St. Jean Baptiste 6, Sorel 6, St. Hyacinthe 5.
New Brunswick	27	321,233	Fredericton 6	St. John river	St. John 26, Portland 15, Moncton 5.
Nova Scotia and Cape Breton	20	440,572	Halifax 36	Atlantic, or Halifax bay	None over 5, Dartmouth 4, Windsor 3, Truro 4, Pictou 4.
Manitoba	123	65,954	Winnipeg 7	Red and Assiniboine rivers	None over 5, Emerson, Brandon, Gimli, Selkirk, Portage-la-Prairie.
British Columbia	341	49,459	Victoria 5	Juan de Fuca str.	None over 5, Nanaimo, Esquimalt.
Prince Edward I'd	2	108,891	Charlottetown 11	Hillsboro bay	None over 5, Georgetown, Summerside.

Ontario, formerly **Upper Canada,** *Quebec,* formerly **Lower Canada,** *New Brunswick,* and *Nova Scotia,* which includes Cape Breton **island, were united under the name, "** *The Dominion of Canada,*" on **July 1st, 1867.** In 1870 *Manitoba* was admitted into the union, in 1871 *British Columbia,* and in 1873 *Prince Edward Island.* The Territories were acquired in 1870, but were organized in May, 1882.

TERRITORIES.

Name.	Area in thousands of sq. miles.	Places of Importance.	Boundaries.
Assiniboia	95	Regina, Qu'appelle, Livingstone, Chesterfield, Wood Mountain, Forts Walsh and Pelly.	*South,* United States, or 49th parallel; *West,* Alberta; *East,* Manitoba; *North,* Saskatchewan, or 52nd parallel.
Saskatchewan	114	Battleford, Cumberland, Prince Albert, Saskatchewan, Forts Pitt, Carlton, and LaCorne.	*North,* Assiniboia; *West,* Alberta; *East,* Manitoba, Lake Winnipeg, and Nelson river; *North,* North-West Territory, or 55th parallel.
Alberta	100	Fort McLeod, Edmonton, Victoria, Rocky Mountain Ho., Forts Calgary, Saskatchewan, Old Bow, and Assiniboine.	*South,* United States; *West,* British Columbia, or Rocky Mountains; *East,* Assiniboia and Saskatchewan; *North,* Athabasca.
Athabasca	122	Dunvegan, Vermillion, Peace River, Athabasca, Forts McLeod and Lesser Slave.	*South,* Alberta; *West,* British Columbia; *East,* North-West Territory, Athabasca river and lake, and Slave river; *North,* North-West Territory, or 60th parallel.
Keewatin	360	Fort York, Norway Ho., Oxford Ho.	*South,* Manitoba; *West,* North-West Territory; *East,* North Territory and Hudson bay; *North,* Arctic Ocean.

The remaining territories, unofficially named *North-West, North,* and *North-East,* cover about one-half of Canada. Population, including Indians, of all the Territories, 57,000. Of the above, *Assiniboia, Saskatchewan* and *Alberta* are being rapidly settled. As yet (1883) all are governed by officers appointed by the Privy Council of Canada.

Ontario.

Capital, Toronto.

<table>
<tr><td rowspan="4">BOUNDARIES.</td><td>*North-east:*—Ottawa river, lake Temiscamingue, and the line due north from this lake to James bay ; or, Quebec and North-East territory.</td></tr>
<tr><td>*North-west:*—James bay, Albany and English rivers, and Lonely lake ; or, Manitoba and North territory.</td></tr>
<tr><td>*South-west:*—Lakes of the Woods, Rainy, Superior, Huron, and St. Clair, and rivers Rainy, Pigeon, St. Mary, St. Clair, and Detroit ; or, Minnesota and the *great lakes.*</td></tr>
<tr><td>*South-east:*—Lakes Erie and Ontario, and rivers Niagara and St. Lawrence.</td></tr>
</table>

GOVERNMENT.

1. *Legislative:*—Legislative Assembly, consisting of :—

    (a) Eighty-eight members, *elected* by the people of Ontario.

    (b) A Lieutenant-Governor, *appointed* by the Privy Council of Canada.

2. *Executive:*—A Council, composed of the Lieutenant-Governor and the leaders of the ruling party in the Assembly. These must be members of the Assembly.

    The *revenue* is derived chiefly from :—The Dominion of Canada, the sale of Crown lands and of timber limits, interest, licenses, law stamps, &c.

    The *expenditure* is chiefly on account of :—Civil government, justice, asylum and other institutions, education, railways, &c.

Lakes.

On the boundaries:—Lonely, Woods, Rainy, Nameukan, Superior, Huron, St. Clair, Erie, Ontario, Thousand Islands, St. Francis, Temiscamingue, and Abittibi.

Within the boundaries:—Nipigon, Shebandowan, Nipissing, Conchiching, Simcoe, Muskoka, Rosseau, Joseph, the Victoria and Peterborough chain (Balsam, Sturgeon, Pigeon, Buckhorn, Chemong or Mud, Stoney or Trout or Clear, Scugog, Rice), Rideau, and Mississippi.

Rivers.

On the boundaries:—Albany, English, Winnipeg, Rainy, Pigeon, St. Mary, St. Clair, Detroit, Niagara, St. Lawrence, and Ottawa.

Into lake Superior:—Pigeon, Kaministiquia, Nipigon, and Michipicoten.

Into lake Huron:—Sable, Saugeen, Maitland, Bayfield, Aux Sables, and St. Mary.

Into Georgian bay:—Mississagua, Spanish, French, Magnetewan, Muskoka, Severn, Nottawasaga, and Sydenham.

Into St. Clair:—Sydenham (Bear Creek), Thames, and St. Clair.

Into lake Erie:—Grand, Detroit, Kettle, and Otter.

Into lake Ontario and bay of Quinte:—Niagara, Credit, Etobicoke, Humber, Don, Rouge, Trent, Moira, Salmon or Shannon, and Napanee.

Into the St. Lawrence:—Raisin, and Beaudette.

Into the Ottawa:—Nation, Rideau, Mississippi, Madawaska, Bonnechere, Muskrat and Indian, Petawawa, Mattawan, and Montreal.

Canals. Rideau, between Kingston and Ottawa, passing *Newboro*, *Smith's Falls*, and *Merrickville*. *Note.*—A branch runs from Smith's Falls to Perth.

Welland, between lakes Erie and Ontario, passing *Port Colborne*, *Welland*, *Port Robinson*, *Allanburgh*, *Thorold*, *Merritton*, *St. Catharines*, and *Port Dalhousie*.

St. Lawrence, to overcome the various rapids on the St. Lawrence, passing *Iroquois*, *Morrisburgh*, *Dickinson's Landing*, and *Cornwall*.

Note 1.—The chain of lakes in Victoria and Peterboro counties is rendered navigable by *locks* at *Bobcaygeon*, *Lindsay*, *Rosedale*, *Young's Point* near *Lakefield*), and *Peterboro*. Steamers run between Port Perry, on lake Scugog, and Bridgenorth, on Chemong lake, within four miles of Peterboro. The Trent Valley canal will pass through this chain.

2.—Lakes Superior and Huron are joined by a canal, the Sault Ste. Marie, on the United States shore.

3.—The St. Clair Flats canal is simply a dredged channel in lake St. Clair.

Islands. Silver and Michipicoten, in Superior; Joseph and Manitoulin, in Huron; Pelee and Long Point, in Erie; Amherst, Wolfe, and Garden, in Ontario; and Thousand, in St. Lawrence. Drummond, in Huron; Grand and Goat, in the Niagara; and the Ducks, in **Ontario, belong** to the United States; **Allumette and Calumet,** in the Ottawa, **belong to Quebec.**

Bays and Gulfs. Thunder, Black, Nipigon, and Michipicoten, in lake Superior; Manitou and Georgian, with its ports—**Parry Sound,** Matchedash, Nottawasaga, Owen Sound, and Colpoy—in lake Huron; Rondeau, Long Point, and Maitland, in lake Erie; **and** Burlington, **Toronto,** Presqu'Isle, Wellers Wellington, South, and Quinte, in lake Ontario.

An *incorporated* village is supposed to contain............................ 800 to 2,000 inhabitants.

A town   "   " ....2,000 to 9,000 "

A city   "   " .... above 9,000 "

Note.—Charters of incorporation are granted by the Legislative Assembly, and confer privileges in the management of local affairs. Many villages, not incorporated, are larger than some incorporated.

A township is a subdivision of a county, surveyed into lots or farms.

A county is a division of a province, for political, judicial, educational, and local improvement purposes.

A province, in Canada, is a division of the country, with power:—To amend its laws; to manage and to sell its public lands and timber; to establish and to maintain public reformatories, prisons, hospitals, charities, &c.; to control its municipal institutions; to manage tavern licenses for revenue purposes; to administer justice; to direct its educational interests, &c.

Note.—The authority of the Parliament of **Canada** extends to:—The public debt, trade and commerce, raising money by any system of taxation, postal service, census, militia, navigation, currency and coinage, banking, weights and measures, criminal law, marriage and divorce, &c., and all classes of subjects not *expressly* stated as coming under the authority of the provinces.

Counties of Ontario.

Towns are in black-faced type, incorporated villages in *italics*, and other important villages in Roman. The figures indicate *in hundreds* the population.

Note.—Unless otherwise stated, the *first named* is the "*county town,*" *i. e.*, where the County Courts sit, the County Council meets, the county buildings (gaol, court-house, etc.) are.

ON THE GEORGIAN BAY.

Counties.	Cities, towns, and villages.
Simcoe	**Barrie** 50, **Collingwood** 42, **Orillia** 30, *Bradford* 12, *Alliston* 11, *Midland* 11, *Penetanguishene* 11, *Stayner* 10, *Sunnidale*, Allandale, Angus, Beeton, Belle Ewart, Bondhead.
Grey	**Owen Sound** 44, **Meaford** 19, *Durham* 11, *Mt. Forest* in part, *Wiarton* in part, Flesherton, Hanover.
Bruce	**Walkerton** 44, **Kincardine** 29, *Port Elgin* 14, *Lucknow* 12, *Paisley* 12, *Southampton (Saugeen)* 12, *Chesley* 9, *Teeswater* 9, *Wiarton* 8, *Tara*, *Tiverton*.

ON LAKE HURON.

Counties.	Cities, towns, and villages.
Bruce	See above.
Huron	**Goderich** 46, **Clinton** 26, **Seaforth** 25, **Wingham** 20, *Exeter* 18, *Blyth* 9, *Bayfield* 7, *Wroxeter* 6, *Brussels* 6, Lucknow (in part).
Lambton	**Sarnia** 39, **Petrolia** 35, *Forest* 17, *Point Edward* 13, *Watford* 12, *Wyoming* 9, *Alvinston* 8, *Thedford* (Widder Stn.) 7, *Arkona* 6, *Oil Springs* 6, Courtright.

ON LAKE ERIE.

Counties	Cities, towns, and villages.
Essex	**Sandwich** 12, **Windsor** 67, **Amherstburg** 27, *Leamington* 15, *Kingsville* 9, *Belle River* 6, Colchester, Essex Centre.
Kent	**Chatham** 60, **Dresden** 20, **Ridgetown** 16, *Wallaceburg* 16, *Blenheim* 13, *Rothwell* 10, *Thamesville* 8, Rondeau (Shrewsbury).
Elgin	**St. Thomas** 90, *Aylmer* 16, *Port Stanley* 7, *Springfield* 6, *Vienna* 6, *Dunwich* (Wallacetown), Port Burwell.
Norfolk	**Simcoe** 27, *Port Dover* 12, *Waterford* 11, Port Rowan, Port Ryerse.
Haldimand	**Cayuga** 29, *Dunnville* 18, *Caledonia* 13, Jarvis, Hagersville, York, Canfield, Canboro.
Welland	**Welland** 19, **Thorold** 25, **Niagara Falls** (Clifton) 24, *Port Colborne* 18, *Chippawa* 7, *Fort Erie* 7, Drummondville, Ridgeway, Allanburgh.

ON LAKE ONTARIO AND BAY OF QUINTE.

Counties.	Cities, towns, and villages.
Lincoln	**St. Catharines** 100, **Niagara** 15, *Merritton* 18, *Port Dalhousie* 12, *Grimsby* 7, *Beamsville* 7, Smithville, Queenston.
Wentworth	**Hamilton** 300, **Dundas** 87, *Waterdown*, Ancaster, Stoney Creek.
Halton	**Milton** 13, **Oakville** 17, *Georgetown* 15, *Burlington* 11, *Acton* 9, Bronte, Norval.
Peel	**Brampton** 30, *Streetsville* 8, *Bolton* (Albion) 6.

ON LAKE ONTARIO AND BAY OF QUINTE (CONTINUED).

Counties.	Cities, towns, and villages.
York	Toronto 870, *Yorkville* 50 (about being annexed to Toronto), **Newmarket** 20, *Aurora* 16, *Parkdale* 12, *Markham* 10, *Stouffville* 9, *Richmond Hill* 9, *Weston* 9, *Brockton* 8, *Holland Landing* 6, Scarboro, Thornhill, Georgina (Sutton).
Ontario	Whitby 32, Oshawa 40, *Uxbridge* 19, *Port Perry* 18, **Cannington** 10, Beaverton, Pickering (Duffin's Creek), Brooklin, Prince Albert.
Durham	Port Hope 56, Bowmanville 26, *Millbrook* 12, *Newcastle* 11, Orono, Bethany, Hampton, Cartwright (Williamsburg). *Note.*—Durham and Northumberland are united ; County town, Cobourg.
Northumberland	Cobourg 50, **Campbellford** 15, *Brighton* 15, *Colborne* 11, *Hastings* (in part), Baltimore, Harwood.
P. Edward	Picton 30, *Wellington* 6, *Hallowell (Bloomfield)*, Consecon.
Hastings	Belleville 66, **Trenton** 31, *Deseronto (Mill Point)* 17, Madoc 11, *Stirling* 9, *Wollaston*, Marmora.
Lennox	Napanee 37, Adolphustown. *Note.*—Lennox and Addington are united ; County town, Napanee.
Addington	*Newburgh* 14, *Bath* 6, Odessa, Tamworth.
Frontenac	Kingston 140, *Portsmouth* 18, *Garden Island* 5, Loughboro (Sydenham).

ON THE RIVER ST. LAWRENCE.

Counties.	Cities, towns, and villages.
Leeds	**Brockville** 76, *Gananoque* 29, *Newboro* 5, Farmersville. *Note.*—Leeds and Grenville are united ; County town, Brockville.
Grenville	**Prescott** 30, *Kemptville* 12, *Merrickville* 9, *Cardinal*.
Dundas	*Morrisburgh* 18, *Iroquois* 10. *Note.*—Dundas, Stormont, and Glengarry are united ; County town, Cornwall
Stormont	**Cornwall** 45, Dickinson's Landing.
Glengarry	Lancaster, Alexandria, Martintown, Williamstown.

ON THE RIVER OTTAWA.

Counties.	Cities, towns, and villages.
Renfrew	Pembroke 29, *Arnprior* 22, *Renfrew* 16.
Carleton	**Ottawa** 274, *New Edinburgh* 30, *Richmond* 5.
Russell	Russell (Duncanville). *Note.*—Russell and Prescott are united ; County town, L'Orignal.
Prescott	*L'Orignal* 9, *Hawkesbury* 20, *Vankleek Hill*.

NOT ON THE BOUNDARIES.

Counties.	Cities, towns and villages.
Lanark	Perth 25, **Almonte** 27, *Smith's Falls* 21, *Carleton Place* 20, **Lanark** 8, Packenham.
Peterboro	Peterborough 60, *Ashburnham* 13, *Hastings* 9, *Lakefield* 9, *Norwood* 9.
Victoria	Lindsay 51, *Fenelon Falls* 12, **Bobcaygeon** 8, **Omemee** 8, Oakwood, Woodville, Kinmount, Coboconk.
Dufferin	Orangeville 29, *Shelburne* 8, Dundalk.

NOT ON THE BOUNDARIES (Continued).

Counties.	Cities, towns and villages.
Wellington	Guelph 100, Mt. Forest 22, Harriston 18, Fergus 18, **Elora** 14, *Arthur* 13, *Clifford* 8, *Drayton* 6, *Erin*, Palmerston (in part).
Waterloo	Berlin 41, Galt 52, Waterloo 21, *Preston* 17, *New Hamburg* 13, *Hespeler* 7, Ayr 12, Elmira, Breslau, Conestogo 10.
Perth	Stratford 83, St. Mary's 35, Listowell 27, Mitchell 23, Palmerston 19, *Milverton* 6, Shakespeare.
Brant	Brantford 96, Paris 32, Harrisburg.
Oxford	Woodstock 54, Ingersoll 44, Tilsonburg 20, *Norwich* 15, *Embro* 7, Hawtrey, Otterville Drumbo.
Middlesex	London 198, Strathroy 39, *London East* 29, *Petersville* 16, *Parkhill* 16, *Lucan* 10, *Ailsa Craig* 9, *Glencoe* 8, *Wardsville* 6, *Newbury* 6, Westminster, Komoka.
Haliburton	Minden, Haliburton, Snowden.

Districts or unorganized counties.

ON THE GEORGIAN BAY.

Counties.	Towns and Villages.
Muskoka	*Bracebridge* 15, Gravenhurst, Severn Bridge.
Parry Sound	*Parry Sound* 13, Maganetewan.
Algoma	*Sault Ste. Marie* 9, Bruce Mines.

ON LAKE SUPERIOR.

Counties.	Towns and Villages.
Algoma	(See "On the Georgian Bay.")
Thunder Bay	Pr. Arthur's Landing 13, Fort William 7, Rat Portage, Fort Francis, Fort Moose.

Note.—**Nipissing**, containing Mattawa and Hopefield, borders on the Ottawa river.

THE TEN CITIES OF ONTARIO ARE:

Toronto,	86,415.	Guelph,	9,890.
Hamilton,	35,961.	St. Catharines,	9,631.
Ottawa,	27,412.	Brantford,	9,616.
London,	19,746.	Belleville,	9,516.
Kingston,	14,091.	St. Thomas,	8,367.

Note.—Important places with only one railway are in Roman. Junctions or crossings of *two* roads are in *italics.* Junctions or crossings of *more than two* are **black-faced.**

RAILWAYS OF ONTARIO.

Railways.	Miles.	Towns, crossings, junctions, &c.
Grand Trunk (Main Line)	961	*Detroit, *Port Huron, Sarnia (Point Edward), Parkhill, Ailsa Craig, *Lucan, St. Mary's,* Stratford, *Berlin,* Guelph, *Georgetown, Brampton,* Toronto, *Scarboro, Whitby,* Oshawa, Bowmanville, *Port Hope, Cobourg, Trenton,* Belleville, *Deseronto,* Napanee, *Kingston,* Brockville, Prescott, Morrisburg, Cornwall, Lancaster, *Coteau,* *Montreal, *Portland.
Buffalo and Lake Huron Division	160	*Buffalo, Fort Erie, *Port Colborne,* Dunnville, Canfield, *Caledonia,* Brantford, *Paris, Drumbo, Tavistock,* Stratford, Seaforth, *Clinton,* Goderich.
St. Mary's and London Division	22	St. Mary's, London.
Waterloo and Galt Division	15	Waterloo, *Berlin,* Galt.
Lake Erie & Georgian Bay Division	167	Wiarton, Tara, Chesley, Hanover, Harriston, Palmerston, *Listowel,* Milverton, Stratford, *Tavistock,* Woodstock, *Norwich, Hawtrey, Simcoe, Port Dover.*
Palmerston and Durham Branch	26	Palmerston, *Mount Forest,* Durham.
Great Western Division of Grand Trunk	229	Windsor, *Chatham, Glencoe, Komoka,* London, *Ingersoll,* Woodstock, *Paris,* Harrisburg, Dundas, Hamilton, St. Catharines, *Merritton,* Niagara Falls (Clifton).
Wellington, Grey, and Bruce Division of Great Western	137	Brantford, Harrisburg, Galt, Preston, *Hespeler,* Guelph, *Elora, Fergus, Drayton,* Palmerston, Harriston, Clifford, Walkerton, Paisley, Port Elgin, Southampton.
Kincardine Branch	67	Palmerston, *Listowel,* **Brussels, Bluevale,** *Wingham, Lucknow,* Kincardine.
Brantford, **Norfolk,** and Port Burwell Division	35	(Harrisburg 8 miles), **Brantford,** *Norwich, Tilsonburg.*
Toronto Division	39	Hamilton, **Waterdown, Burlington,** Oakville, **Toronto.**
London, Huron, & Bruce Division	74	London, *Hyde Park, Lucan,* **Exeter, Clinton,** Blyth, *Wingham* (Kincardine 25 miles).
London and Port Stanley Division	24	London, Westminster, **St. Thomas,** Port Stanley.
Loop, or Air Line, Division	145	*Glencoe,* St. Thomas, Aylmer, Tilsonburg, *Simcoe, Jarvis,* Cayuga, Welland Junction, Fort Erie, *Buffalo.
Niagara Falls Branch	17	Welland Junction, Welland, Port Robinson, *Allanburgh,* Stamford, Niagara Falls.
Sarnia Branch	51	(London 10 miles), *Komoka,* Strathroy, Watford, *Wyoming,* Sarnia.
Petrolia Branch	6	*Wyoming, Petrolia.*
Toronto, Grey, and Bruce	122	Toronto, Woodbridge, *Cardwell* (Caledon East), *Alton,* Orangeville, Shelburne, Flesherton, Owen Sound.
Teeswater Branch	70	Orangeville, Arthur, *Mount Forest,* Harriston, Wroxeter, Teeswater.
Midland	120	*Port Hope, Millbrook,* Bethany, *Omemee,* Lindsay, *Lorneville,* Beaverton, *Atherley, Orillia,* Waubaushene, Midland.
Peterboro Branch	22	*Millbrook,* Peterboro, **Lakefield.**
Whitby and Haliburton Division,	101	*Whitby,* **Brooklin,** Prince Albert, Port Perry, **Manilla,** Lindsay, Fenelon Falls, Kinmount, Haliburton.
Nipissing Division	88	Toronto, *Scarboro, Stouffville,* Uxbridge, *Wick,* Sunderland, *Lorneville,* Coboconk.

* Not in Ontario.

RAILWAYS OF ONTARIO.—Continued.

Railways.	Miles.	Towns, crossings, junctions, &c.
Sutton Branch	27	*Stouffville*, Sutton, Jackson's Point.
GRAND JUNCTION DIVISION	66	Peterboro, Keene, Hastings, Campbellford, **Stirling, N. Hastings Junction, Belleville**.
Madoc Branch	15	*N. Hastings Junction, Madoc.*
TORONTO AND OTTAWA DIVISION	..	Toronto, *Nip. Div.* to *Wick, Manilla, W. & H. Div.* to Lindsay, *Main Line* to *Omemee*, Peterboro, *Grand Jn. Div.* to Madoc, Bridgewater, *Perth?* Ottawa? or *Cornwall?*
Northern & North-Western	151	*Port Dover, Jarvis, Hagersville, Caledonia*, Hamilton, *Burlington, Milton, Georgetown, Riverdale, Caledon East* (Cardwell Jn.), *Beeton*, Alliston, *Collingwood.*
NORTHERN DIVISION	115	Toronto, Thornhill, Aurora, Newmarket, Holland Landing, Bradford, Allandale, **Barrie, Colwell**, Angus, Stayner, Collingwood, Meaford.
Muskoka Branch	51	Allandale, Barrie, *Orillia*, **Atherley**, Gravenhurst.
Barrie and Beeton Branch	26	Barrie, Allandale, **Thornton**, Cookstown, *Beeton.*
North Simcoe Branch	30	Barrie, **Allandale**, *Colwell*, Penetanguishene.
Credit Valley	121	Toronto, *Streetsville*, **Milton, Galt**, Ayr, *Drumbo*, **Woodstock**, *Ingersoll*, **St. Thomas.**
Orangeville Branch	34	*Streetsville, Brampton, Riverdale, Church Falls, Alton,* Orangeville.
Elora Branch	27	*Church Falls*, Erin, *Fergus, Elora.*
Canada Southern	228	*Buffalo, Victoria (Ft. Erie), Niagara Jn.,* Welland Jn., Canfield, *Hagersville*, Waterford, *Hawtrey*, Tilsonburg, St. Thomas, *St. Clair Jn., Ridgetown, Charing Cross, Essex Centre*, Maidstone, Sandwich, Windsor, *Detroit.*
Amherstburg Branch	16	*Essex Centre*, **Colchester Station**, Amherstburg.
ST. CLAIR DIVISION	66	St. Thomas, *St. Clair Jn., Loop Line Jn.*, Delaware, *Gt. W. Jn.,* Alvinston, Oil City, *Petrolia Jn.*, Courtright
Petrolia Branch	7	*Petrolia Jn.* (near Oil City, and also Oil Springs), *Petrolia*
NIAGARA DIVISION	30	*Buffalo, Victoria (Fort Erie), Niagara Jn.,* **Chippawa,** Niagara Falls, Queenston, Niagara.
Canada Pacific	199	Ottawa, *Carleton Place Jn.*, Almonte, Packenham, Arnprior, Renfrew, Pembroke, Petawawa, Mattawa.
BROCKVILLE DIVISION	46	*Brockville, Smith's Falls, Carleton Place* (Ottawa 29 miles).
Perth Branch	12	*Smith's Falls, Perth.*
St. Lawrence and Ottawa	54	*Prescott,* Spencerville, **Kemptville,** *Chaudière Jn.,* **Ottawa.**
Canada Atlantic	..	Ottawa, *Chaudière Jn.,* Alexandria, *Coteau Landing.*
Welland	25	*Port Colborne, G. W. Jn., C. S Jn.,* Welland, Port Robinson, *Allanburgh,* **Thorold**, *Merritton*, **St. Catharines,** Port Dalhousie.
Kingston and Pembroke	61	*Kingston*, Sharbot Lake, Mississippi.
Central Ontario	32	*Trenton*, Consecon, Wellington, Bloomfield, Picton.
Cobourg and Rice Lake (Cobourg, Peterboro, & Marmora)	15	*Cobourg,* **Baltimore**, Harwood (on Rice lake).
†*Huron and Erie*	..	Rondeau, Blenheim, *Charing Cross, Chatham*, Dresden, Oil Springs. (Branch from Dresden to Wallaceburg.)
†*Ontario and Quebec*	..	Toronto, *Markham, Bethany,* Peterboro, **Madoc, Perth,** Ottawa.

† Under construction (1883). * Not in Ontario.

Note.—The *Ontario Central* is extending up the Trent valley north of *Trenton.* There is a short line from the Grand Trunk to *Deseronto* on the Bay of Quinte. The *Napanee and Tamworth* is graded. The Ontario & Sault Ste. Marie Railway is being rapidly pushed forward to run in connection with the Midland.

Rivers.

Saugeen :—Southampton, Paisley, Walkerton, Teeswater, Chesley, Hanover, Durham, Clifford, Mount Forest.

Maitland :—Goderich, Wingham.

    North branch :—Wingham, Wroxeter, Gorrie, Fordwich, Harriston.

    Little Maitland :—Wingham, Bluevale, Palmerston.

    South branch :—Brussels, Cranbrook, Trowbridge, Listowell.

Bayfield :—Bayfield, Clinton, Seaforth.

Aux Sables :—Arkona, Parkhill, Ailsa Craig, **Exeter, Lucan.**

Sydenham, or Bear Creek :—Wallaceburg.

    North branch :—Wallaceburg, Oil Springs, Petrolia.

    South branch :—Wallaceburg, Dresden, Florence, Alvinston, Watford, Strathroy.

Thames :—Chatham, Thamesville, Bothwell, Wardsville, Komoka, London.

    North branch :—London, St. Mary's, Mitchell.

        Avon :—Stratford Shakespeare.

    South branch :—London, Ingersoll, Beachville, Woodstock.

Otter :—Port Burwell, Vienna, Tilsonburg, Otterville, Norwich.

Grand :—Port Maitland, Dunnville, Cayuga, York, Caledonia, Brantford, Paris, Galt
    Preston, Breslau, Conestogo, Elora, Fergus.

    Nith :—**Paris, Ayr, New Hamburg, Baden, Wellesley, Millbank.**

    Speed :—**Preston, Hespeler, Guelph, Erin.**

    Conestogo :—**Conestogo, Glenallan, Drayton, Arthur.**

Credit :—Port Credit, Streetsville, **Orangeville.**

Humber :—Lambton Mills, Weston, Woodbridge, Kleinburg, Nobleton, Bolton.

Holland :—Newmarket, Aurora, Holland Landing, Bradford.

Trent :—Trenton, Frankford, Campbellford, Hastings.

    Otonabee :—Peterboro, Ashburnham, Lakefield.

    Scugog :—Lindsay, Port Perry (on lake Scugog).

    Gull :—Coboconk, Minden.

        Note.—Fenelon Falls is between Sturgeon and Scugog lakes ; Bobcaygeon, between Sturgeon and Pigeon lakes ; Omemee is on Pigeon river.

Moira :—Belleville, Bridgewater, Tweed.

Raisin :—Lancaster, Williamstown, Martintown.

Rideau :—Ottawa, Merrickville, Smith's Falls. Perth is on the Tay.

Mississippi :—Packenham, Almonte, Carleton Place, Lanark.

Ports of Ontario

Lake Superior :—Michipicoten, Nipigon, Fort William, **Prince Arthur's Landing.**

Georgian Bay :—**Parry Sound, Maganetewan, Midland, Penetanguishene, Collingwood, Meaford, Owen Sound, Wiarton (Colpoy Bay).**

Lake Huron :—Sarnia, Bayfield, Goderich, Kincardine, Elgin, Southampton, Bruce Mines, Little Current, Manitowaning.

Lake Erie :—Rondeau, Stanley, Burwell, Eyeau, Rawan, Dover, Maitland, Selborne.

Ports of Ontario

Lake Ontario and Bay of Quinte :—Niagara, Dalhousie, Hamilton, Oakville, Credit, Toronto, Liverpool, Whitby, Oshawa, Bowmanville (Darlington), Port Hope, Presqu'Isle (Brighton), Trenton, Belleville, Deseronto (Mill Point), Napanee, Picton, Kingston.

St. Lawrence :—Gananoque, Brockville, Prescott, Dickinson's Landing, Cornwall.

Industries of Ontario.

Farming :—All except the parts north of lakes Huron and Superior is well adapted for farming.

Lumbering :—Conducted on a large scale in the Georgian Bay and Ottawa and Trent river districts.

Mining :—Iron in Victoria, Peterboro, Hastings, Frontenac, Lanark, &c. ; gold, silver, copper, asbestos, lead, &c., in abundance, north of lakes Huron and Superior, and eastern Ontario. Petroleum abounds in Lambton, and salt in Huron.

Manufacturing :—Carried on extensively in the cities and larger towns.

Fishing and *Shipping* employ a large number of men, chiefly in lakes Superior and Huron.

Railways :—A large number of men also find employment in connection with the railways in Ontario.

Quebec.

Capital, Quebec.

BOUNDARIES.

North :—North-East territory and Labrador.

South :—United States and New Brunswick.

East :—Gulf of St. Lawrence.

West :—Ontario, or the Ottawa river.

Government.

A Legislative Assembly of sixty-five members, *elected* by the people of Quebec.

A Legislative Council of twenty-four members, *appointed* for life by the Lieutenant-Governor of the province.

A Lieutenant-Governor, *appointed* by the Privy Council of Canada.

Lakes.

South of the St. Lawrence :—Champlain (in part), Memphramagog, St. Francis, Megantic, Temiscouata, Matapedia.

North of the St. Lawrence :—Manouan, St. John.

Expansions of the Ottawa :—Temiscamingue, Chat, Two Mountains.

Expansions of the St. Lawrence :—St. Francis, St. Louis, St. Peter

Rivers.

Ottawa :—Rouge, Nation, Lièvre, Gatineau, Coulonge, Moine.

St. Lawrence from the North :—Assomption, St. Maurice, Batiscan, St. Anne, Jacques Cartier, St. Charles, Montmorency, Saguenay.

St. Lawrence from the South :—Chateauguay, Richelieu, Yamaska, St. Francis, Nicolet, Becancour, Chaudière, Etchemin.

The *Temiscouata* flows into the St. John, and the *Matapedia* into the Restigouche.

On the boundaries :—Ottawa, St. John, Restigouche.

Canals. | *Lachine*, from Montreal to Lachine, on the St. Lawrence.

St. Ann's, a lock at the western extremity of Montreal island, to enter the Ottawa from lake St. Louis.

Lake St. Peter requires dredging annually, to preserve *twenty-five* feet of water.

Islands. | *In the Ottawa :*—Allumette and Calumet.

At the junction of the Ottawa and St. Lawrence :—Montreal, Perrot, Jesus (Laval).

Below Quebec :—Orleans, Bic.

In St. Lawrence Gulf :—Anticosti and Magdalens.

Bays and Gulfs. | Murray (Malbaie), Cacouna, Rimouski, Gaspé, Chaleur.

Industries. | Farming, lumbering, mining, shipbuilding, fishing, and manufacturing.

TOWNS OF OVER FIVE THOUSAND INHABITANTS.

Towns.	Population.	Towns.	Population.
Montreal	140,747	Hull	6,890
Quebec	62,446	St. Henri*	6,415
Three Rivers	8,670	St. Jean Baptiste*	5,874
Levis	7,597	Sorel	5,791
Sherbrooke	7,227	St. Hyacinthe	5,321

TOWNS OF OVER TWO THOUSAND INHABITANTS.

Names.	Pop.	Names.	Pop.	Names.	Pop.
St. Cunegonde*	4,549	Valleyfield	3,906	Lachine	2,400
St. Gabriel*	4,500	Laur	3,556	Longueuil	2,355
St. Jean (St. Johns)	4,314	Joliette	3,268	Fraserville	2,291
Hochelaga*	4,111	Coaticook	2,632	Berthier	2,156

* Suburbs of Montreal.

Other places of note are Rimouski, Father Point, Cacouna, Richmond, Lennoxville, Actonvale, Iberville, Farnham, Chambly, Laprairie, Beauharnois, Coteau, Buckingham, Aylmer, Gatineau, and Stanstead.

Note.—The Grand Trunk and the North Shore Railways connect Quebec with Montreal. The Canada Pacific and the Canada Atlantic join Montreal and Ottawa. The Grand Trunk connects Montreal with Ontario, whilst several other roads run east and south-east to the states along the border. Sherbrooke is a railway centre of importance.

New Brunswick.

Capital, Fredericton.

BOUNDARIES.	*North :—*Quebec, or bay of Chaleur and river Restigouche. *East :—*St. Lawrence gulf. *South :—*Bay of Fundy. *West :—*Maine, and the St. Croix river.
Government.	A Legislative Assembly, *elected* by the people ; a Legislative Council, *appointed* by the Lieutenant-Governor ; and a Lieutenant-Governor, *appointed* by the Privy Council of Canada. An Executive Council of the Assembly advises the Lieut.-Governor.
Rivers.	*On the boundaries :—*Restigouche, St. John, St. Croix. *Into the Gulf of St. Lawrence :—*Miramichi, Richibucto. *Into the Bay of Fundy :—*St. John, Kennebeccasis, Petitcodiac.
Islands.	Shippegan, Miscou, Campobello, and Grand Manan.
Bays and Gulfs.	Chaleur, Miramichi, Richibucto, Shediac, Verte, Cumberland, Shepody, Chignecto, Fundy, St. John, Passamaquoddy.
Industries.	Lumbering, shipbuilding, fishing, farming, mining.

TOWNS.

Name.	Pop.	Name.	Pop.	Name.	Pop.	Name.	Pop.
St. John	26,127	Newcastle	4,000	Milltown	1,664	Carleton	4,000
Portland	15,226	St. George	3,412	Shediac	1,500	Quaco	....
Fredericton	5,218	St. Stephen	2,338	Sackville	1,500	Dorchester	....
Moncton	5,032	Woodstock	2,487	Richibucto	1,200	Bathurst	....
Chatham*	4,600	St. Andrews	2,228	Dalhousie	1,000	Shippegan	....

* Chatham generally goes by the name—Miramichi.

Note 1.—The St. John is navigable to Woodstock. There is only one lake—Grand—in New Brunswick ; and one—Grand—on the Maine border.

   2.—Shipbuilding is extensively carried on along the shores of the bay of Fundy, and in the Miramichi district.

   3.—The Intercolonial Railway connects *St John* with Moncton, Shediac, Sackville, Newcastle, Bathurst, Dalhousie, Halifax, Quebec, etc.

     St. Stephen and St. Andrew are on the New Brunswick and Canada Railway, which runs near Woodstock, through Maine, to Quebec province.

     Fredericton is on the New Brunswick Railway from St. John up the St. John valley to the northern part of the province.

Nova Scotia.

Capital, Halifax.

BOUNDARIES.
- *North:*—Northumberland straits and gulf of St. Lawrence.
- *South-East:*—Atlantic Ocean.
- *West:*—Bay of Fundy, Chignecto bay, and New Brunswick.

Government. The same form as New Brunswick has.
Islands. Cape Breton, Sable, Long.
Industries. Fishing, shipbuilding, lumbering, farming, mining.

BAYS AND GULFS.

On the Atlantic sea-board.	On the northern coast.	On the western coast.
Chedabucto, Ship, Halifax, Margaret, Mahone, Liverpool, Shelbourne, Yarmouth.	Verte, Pugwash, Pictou, Antigonish, St. George, Bras d'or, Sydney.	Fundy, St. Mary, Annapolis (Port Royal), Mines, Cobequid, Avon, Chignecto, Cumberland,

TOWNS.

On the Atlantic coast.	Pop.	On the northern coast.	Pop.	On the western coast.	Pop.
Halifax,	36,100	Sydney,	3,600	Truro,	3,461
Dartmouth,	3,786	Pictou,	3,403	Windsor,	3,019
Yarmouth,	4,000	Millsville,	3,340	Scotch Village,	2,811
Lunenburg,	3,000	New Glasgow,	2,800	Annapolis,	2,800
Shelburne,	2,000	Antigonish,	2,000	Amherst,	2,000
Liverpool,	1,500	Pugwash,	1,000	Digby,	1,870
Guysboro,	1,000			Marshall,	1,077
Arichat,	1,000			Bridgetown,	1,400
				Londonderry Stn.	1,000

Note 1.—**Louisburg**, on Cape Breton Island, was taken from the French by Amherst, Wolfe, and Boscawen in 1758.

2.—The **Gut of Canso** connects Northumberland straits with the Atlantic, and separates Cape Breton from the mainland.

3.—The **Intercolonial Railway** connects Halifax with Truro, Pictou, New Glasgow, Amherst, St. John, Quebec, etc. The Windsor and Annapolis Railway runs from Halifax to Annapolis via Windsor, Bridgetown, etc.

Prince Edward Island.
Capital, Charlottetown.

BOUNDARIES.	*North, East,* and *West* :—Gulf of St. Lawrence. *South* :—Northumberland straits.
Government.	The same form as New Brunswick has.
Bays and Gulfs.	Hillsboro, Cardigan, Bedeque, Richmond, Murray.
Towns, &c.	Charlottetown, 11,484; Summerside, 2,853; Georgetown, 1,118; *Alberton, Cardigan,* Tignish, Souris East.
Industries.	Fishing, farming, shipbuilding.

Note 1.—The *Prince Edward Island* Railway runs from Tignish, at the western, to Souris East, the eastern extremity of the island. One branch runs to Charlottetown, and another to Georgetown.
2.—Ferries run to Shediac, Pictou, New Glasgow, &c., from Summerside, Charlottetown, and Georgetown.

British Columbia.
Capital, *Victoria.*

BOUNDARIES.	*North* :—North-West territory, or 60th parallel. *South* :—United States, or 49th parallel. *East* :—Alberta and Athabasca territories. *West* :—Pacific Ocean and Alaska.
Government.	The same form as Ontario has.
Lakes.	Okanagan, Arrow, Kootenay, François, Quesnel.
Rivers.	Fraser, Columbia, Kootenay, Quesnel, Thompson, Skeena, Findlay, Parsnip, Peace, Athabasca, Okanagan.
Islands.	Vancouver, Queen Charlotte, Scotts, San Juan archipelago.
Bays and Gulfs.	Georgia, Burrard, Bute, Jervis, Nepean, Portland, Nootka, Barclay.
Straits, Sounds.	Queen Charlotte, Dixon, Johnston, Broughton, Juan de Fuca.
Cities, &c.	Victoria, 5,925 ; New Westminster, Esquimalt, Nanaimo, Yale, Cassiar, Kamloops.
Industries.	Mining, fishing, farming.

Note.—The chief passes through the Rocky Mountains are Kootenay, Crow Nest, Kananaskis, Vermillion, Kicking Horse, Athabasca, Yellowhead, Pine River, Peace River.

Newfoundland
Capital, St. John's.

BOUNDARIES.	Atlantic Ocean, gulf of St. Lawrence, straits of Belle Isle.
Government.	Same form as New Brunswick has, but the Governor is appointed by the Privy Council of Great Britain and Ireland. N.B.—Newfoundland is not yet in the Dominion of Canada.

Note.—Lakes, rivers, capes, islands, bays, &c., abound.

Cities, Towns, &c.	St. John's, 30,000, on St. John's harbor ; Harbor Grace, 7,500, and Carbonear, on Conception bay ; Heart's Content, on Trinity bay ; and Placentia, on Placentia bay, are the most important towns.
Industries.	Fishing, of, cod herring, salmon, and seal.

N. B.—The coast of Labrador is under the government of Newfoundland.

United States.

Capital, Washington.

Area, 3,100,000 miles.   Population, 50,152,866

BOUNDARIES.
{
North :—Canada and the great lakes.
South :—Gulf of Mexico and Mexico.
East :—Atlantic Ocean.
West :—Pacific Ocean
}

Government.   1. *Legislative :* A Congress, consisting of :—

 (a) Senate, whose members are *chosen* for six years by the *State Legislatures,* two being sent by each.

 (b) House of Representatives, whose members are *elected* every second year by popular vote. **The number which each state is entitled to send is determined** by the decennial census, **there being one for every 135,230.**

 The President has a *veto* power on legislation, but this may be annulled by a **two-thirds majority of** the members of each House.

2. *Judicial :* A Supreme Court, made up of a Chief Justice and eight Justices, *appointed for life* by the President, by and with the consent of the Senate.

3. *Executive :* A President, **assisted by seven heads of departments**, chosen by him, but who must be approved of by the Senate.

Note 1.—**Each** *territory* is entitled to send a *delegate* to the House of Representatives. He may not vote at all, but has the right to debate on questions concerning his own territory.

2.—The *States* possess powers somewhat similar to those held by the Provinces of Canada, with the important exception that all subjects not *expressly mentioned* as coming under the authority of the Federal government belongs exclusively to the *States.* Each state controls its *militia.*

3.—**Mode of Electing the President.**—Each State chooses by popular vote "Electors," equal in number to the Senators and Representatives sent by that State to Congress. **These** "Electors" meet at their respective State capitals on an appointed day—the same day throughout **the** United States—and there vote for President by ballot. The ballots are then sent to **Washington** where the President *of the Senate,* in presence of Congress, counts them. The candidate who has received a *majority of the whole number of electoral votes cast,* is declared President. **If no one has a majority, then, from the three highest on the list, the** *House of Representatives* **elects a President. The Vice-President** is *ex-officio* President of the Senate. In case the President dies the Vice-President becomes President ; after him the *pro. tem.* President of the **Senate, and after him the Speaker of the House** of Representatives, succeeds to the vacant Presidency.

As soon as the "Electors" are chosen it is known who is to be **President, because it is generally** well understood how each "Elector" will vote.

States and Territories.

Note 1.—The *territories* are in italics.

2.—The *state* capital is placed first; cities of over 100,000 are in SMALL CAPITALS; between 100,000 and 50,000, black-faced; from 50,000 to 20,000 in *italics*; under 20,000 in Roman.

ON THE ATLANTIC SEABOARD.

Name.	Cities, towns, &c.	Products.
Maine	Augusta, **Portland**, Bangor, Lewiston, Biddeford, Saco, Calais, Bath, Rockland, Belfast, Brunswick.	Grain, butter, cheese, lumber, wool, fish.
New Hampshire	Concord, *Manchester*, Nashua, Dover, Portsmouth.	Grain, dairy produce, lumber, minerals, manufactures.
Massachusetts	BOSTON, **Lowell**, **Cambridge**, **Worcester**, *Lawrence*, *Lynn*, *Fall River*, *Springfield*, *Salem*, *New Bedford*, *Taunton*, *Gloucester*, *Newburyport*, *Haverhill*, *Newton*, Bradford.	Manufactures, dairy produce, grain, fish, minerals.
Rhode Island	**Providence**, *North Providence*, Newport, Woonsocket, Pawtucket.	Manufactures, farm produce, tobacco, fish.
Connecticut	**Hartford**, **New Haven**, *Bridgeport*, *Norwich*, *Meriden*, Waterbury, New London.	Manufactures, tobacco, grain, and dairy produce.
New York	ALBANY, NEW YORK, BROOKLYN, BUFFALO, ROCHESTER, **Troy**, **Syracuse**, *Utica*, *Poughkeepsie*, *Oswego*, *Auburn*, *Elmira*, Cohoes, Newburg, Binghampton, Lockport, Ogdensburg, West Point.	Grain, wool, butter, cheese, manufactures, minerals.
New Jersey	Trenton, NEWARK, JERSEY CITY, **Paterson**, Hoboken, *Elizabeth*, *Camden*, *New Brunswick*, Princeton, Atlantic City, Cape May, Long Branch.	Fruits, grain, manufactures, minerals, as zinc, &c.
Delaware	Dover, **Wilmington**, *Smyrna*, Newcastle.	Fruits, grain, butter, cheese.
Pennsylvania	*Harrisburg*, PHILADELPHIA, PITTSBURG, **Allegheny City**, *Scranton*, *Reading*, *Lancaster*, *Erie*, *Williamsport*, Wilkesbarre.	Grain, fruit, minerals, as coal, iron, lead, zinc; manufactures.
Maryland	Annapolis, BALTIMORE, Cumberland, Frederick.	Fruits, grain, dairy produce, tobacco, oysters, minerals.
District of Columbia	WASHINGTON, *Georgetown*.	Manufactures.
Virginia	Richmond, *Norfolk*, *Petersburg*, *Alexandria*, Portsmouth, Lynchburg, Winchester, Fredericksburg.	Tobacco, grain, dairy produce, minerals.
North Carolina	Raleigh, *Wilmington*, Newbern, Beaufort, Charlotte, Fayetteville.	Cotton, tobacco, rice, lumber, pitch, resin, &c.
South Carolina	Columbia, **Charleston**, Greenville, Georgetown.	Cotton, tobacco, rice, molasses.
Georgia	*Atlanta*, *Savannah*, *Augusta*, Macon, Columbus.	Cotton, rice, corn, potatoes, lumber.
Florida	Tallahassee, Jacksonville, Key West, Pensacola, Augustine.	Cotton, fruits, molasses, lumber.

ON THE GULF OF MEXICO.

Name.	Cities, towns, &c.	Products.
Florida	See "Atlantic seaboard."	
Alabama	Montgomery, **Mobile**, *Birmingham*, Florence.	Cotton, potatoes, rice, molasses, lumber, minerals.
Mississippi	Jackson, Vicksburg, *Natchez*, **Columbus**, Corinth.	Cotton, molasses, rice, lumber, potatoes.
Louisiana	NEW ORLEANS, Baton Rouge, Shreveport, Carrollton.	Cotton, rice, molasses.
Texas	Austin, *Galveston*, San Antonio, **Houston**, Brownsville.	Cotton, corn, live-stock, minerals.

MAP GEOGRAPHY PRIMER.

ON MEXICO.

Name.	Cities, towns, &c.	Products.
Texas	See " On the Gulf of Mexico."	
New Mexico	Santa Fé, Albuquerque, Mesilla.	Gold, silver, live-stock.
Arizona	Tucson, Prescott, Aubrey, Arizona, Gila.	Gold, silver, live-stock.
California	*Sacramento*, SAN FRANCISCO, *Oakland*, Stockton, San José, Los Angelos, San Diego.	Grain, fruits, gold, silver, coal, lumber, fish.

ON THE PACIFIC SEABOARD.

Name.	Cities, towns, &c.	Products.
California	See above.	
Oregon	Salem, Portland, Astoria, Albany, Oregon City.	Grain, gold, silver, lumber, **fish.**
Washington	Olympia, Walla Walla.	Grain, gold, silver, lumber.

ON THE MISSISSIPPI—WEST BANK.

Name.	Cities, towns, &c.	Products.
Louisiana	See " On the Gulf of Mexico."	
Arkansas	*Little Rock*, Helena, Pine Bluff, Napoleon, Fort Smith.	Cotton, grain, live-stock.
Missouri	Jefferson City, ST. LOUIS, *Kansas City*, *St. Joseph*, Hannibal, Sedalia, St. Charles.	Grain, tobacco, live-stock, iron.
Iowa	Des Moines, *Davenport*, *Dubuque*, *Burlington*, *Council Bluffs*, Keokuk, Clinton, Muscatine, Iowa City, Cedar Rapids.	Grain, live-stock, lead, coal, &c.
Minnesota	*St. Paul*, *Minneapolis*, Winona, Rochester, St. Vincent, Duluth, Red Wing, Moorehead, Hastings, Minnehaha.	Grain, lumber, flour, live-stock.

ON THE MISSISSIPPI—EAST BANK.

Name.	Cities, towns, &c.	Products.
Louisiana	See " On the Gulf of Mexico."	
Mississippi	" "	
Tennessee	*Nashville*, **Memphis**, Knoxville, Jackson, Chattanooga, Murfreesboro.	Tobacco, **cotton**, grain, lumber, live-stock, minerals, hemp.
Kentucky	Frankfort, LOUISVILLE, *Covington*, *Newport*, Lexington, Paducah, Maysville.	Tobacco, corn, **hemp**, lumber, live-stock, petroleum.
Illinois	*Springfield*, CHICAGO, *Quincy*, *Peoria*, Bloomington, Aurora, Rockford, Rock Island, Alton, Galena, Galesburg.	Grain, live-stock, minerals.
Wisconsin	Madison, MILWAUKEE, Fond du Lac, Oshkosh, Racine, Janesville, La Crosse, Green Bay, Sheboygan, Prairie-du-Chien, Superior City.	Grain, live-stock, **lumber**, minerals, **as iron, copper, lead,** &c.

ON CANADA AND THE GREAT LAKES.

Name.	Cities, towns, &c.	Products.
Vermont	Montpelier, *Burlington*, Rutland, St. Albans, Brattleboro.	Grain, dairy produce, minerals, manufactures.
Michigan	Lansing, DETROIT, *Grand Rapids*, Jackson, East Saginaw, Saginaw, Kalamazoo, Ann Arbor, Adrian, Bay City, Port Huron, Grand Haven, Cheboygan.	Grain, lumber, salt, minerals, as iron, copper, &c.

Note.—Maine, New Hampshire, New York, Pennsylvania, Ohio, Wisconsin, Indiana, Illinois, Minnesota, *Dakota, Montana, Idaho,* and *Washington* are described elsewhere.

ON THE OHIO.

Name.	Cities, towns, &c.	Products.
Illinois	See "On the Mississippi—East Bank."	
Indiana	Indianapolis, *Evansville, Fort Wayne, Terre Haute,* New Albany, Lafayette, Logansport, Madison.	Grain, lumber, tobacco, live-stock, minerals.
Ohio	*Columbus,* CINCINNATI, CLEVELAND, *Toledo, Dayton,* Sandusky, Springfield, Hamilton, Portsmouth, Zanesville, Akron.	Grain, lumber, live-stock, coal, lead, petroleum.
West Virginia	*Wheeling,* Parkersburg, Martinsburg, Charleston.	Coal, iron, grain, petroleum, lumber.
Kentucky	See "On the Mississippi—East Bank."	

ON THE ROCKY MOUNTAIN PLATEAUX.

Name.	Cities, towns, &c.	Products.
Kansas	Topeka, *Leavenworth*, Lawrence, Atchison, Fort Scott, Wyandotte.	Grain, **coal, iron, live-stock.**
Nebraska	Lincoln, *Omaha*, Nebraska, Plattsmouth.	Grain, coal, live-stock.
Dakota	Yankton, Pembina.	Grain.
Colorado	*Denver*, Central City, Georgetown, Pueblo.	Gold, **silver,** lead, live-stock, grain.
Wyoming	Cheyenne, Laramie, Benton.	Gold, silver, lead, live-stock, grain.
Montana	Helena, Virginia City.	Grain, **gold.**
Utah	*Salt Lake City*, Ogden, Provo.	Grain, gold, coal, silver.
Nevada	Carson, Virginia, Gold Hill, Hamilton.	Silver, gold, coal.
Idaho	Boisé City, Idaho City.	Gold, grain.

Note 1.—Dakota is about to become a *state*.

2.—Indian Territory is set apart for Indians only.

Water Courses of the United States.

ON THE ATLANTIC SEABOARD.

Note.—Those in italics are on the navigable part of the river.

Courses.	Cities, towns, &c.
Penobscot	*Bangor, Rockland, Belfast.*
Kennebec	*Bath,* Augusta.
Androscog'in	*Brunswick, Lewiston.*
Saco	*Saco, Biddeford.*
Merrimac	*Newburyport, Haverhill, Bradford,* Lawrence, Lowell, Nashua, Manchester, Concord.
Connecticut	*Hartford,* Springfield, Greenfield, Brattleboro.
Hudson	*Brooklyn, New York, Jersey City, Hoboken, West Point, Newburg, Poughkeepsie, Albany, Troy, Cohoes, Saratoga.*
MOHAWK	*Cohoes,* Schenectady, Utica, Rome.
Delaware	*Wilmington, Camden, Philadelphia, Trenton, Easton.*
SCHUYLKILL	*Philadelphia, Reading.*
Chesapeake	*Baltimore, Annapolis, Yorktown.*
SUSQUEHANNA	Harrisburg, Lancaster, Wilkesbarre, Scranton, Elmira, Binghampton.
POTOMAC	*Alexandria, Washington, Georgetown,* Harper's Ferry, Cumberland.
JAMES	*Norfolk, Portsmouth, Richmond,* Lynchburg.
Appomatox	*Petersburg.* Head of tide water navigation.
Savannah	*Savannah, Augusta*

ON THE GULF OF MEXICO.

Courses.	Cities, towns, &c.
Mobile	Mobile.
Mississippi	*New Orleans, Baton Rouge, Natchez, Vicksburg, Napoleon, Memphis, Cairo, St. Louis, Alton, Hannibal, Quincy, Keokuk, Muscatine, Rock Island, Davenport, Clinton, Galena, Dubuque, Prairie-du-Chien, Winona, Hastings, St. Paul, Minneapolis, Crow Wing.*
DES MOINES	*Keokuk, Des Moines.*
MISSOURI	*St. Louis, Jefferson City, Kansas City, Leavenworth, St. Joseph, Nebraska, Council Bluffs, Omaha, Sioux City, Yankton, Fort Benton, Helena*
ARKANSAS	*Napoleon, Little Rock.*
RED	*Shreveport.*
WISCONSIN	*Prairie-du-Chien, Portage.*
ILLINOIS	*Peoria, Ottawa, Joliet, Chicago* (by canal).
OHIO	*Cairo, Paducah, Evansville, New Albany,* Louisville, Covington, Newport, Cincinnati, Wheeling, Alleghany City, Pittsburg.
Grande del Norte	*Brownsville, Rio Grande City,* Santa Fé, Pueblo (Col.), *Matamoras* (in Mexico).

ON THE PACIFIC SEABOARD.

Courses.	Cities, towns, &c.
Colorado (Cal.)	*Fort Yuma, Arizona City, Aubrey,* Georgetown (Col).
GILA	*Arizona City, Gila,* Tucson (on Santa Cruz branch).
Sacramento	*San Francisco* (bay), *Sacramento, Shasta.*
Columbia	*Astoria, Portland.*
WILLAMETTE	*Portland, Oregon, Salem, Albany.*
LEWIS or SNAKE	*Walla Walla, Lewiston,* Idaho City. Boisé City.

PORTS NEAR THE CANADIAN BORDER.

Note.—Those in black-faced type trade largely with Canada.

Waters.	Cities, towns, &c.
St. Lawrence	Ogdensburg, Morristown, Clayton, Cape Vincent.
Ontario	Sackett's Harbor, Watertown, Pulaski, Oswego, Fair Haven, Charlotte (Rochester), Oak Orchard, Youngstown, and Lewiston (the last two are on the Niagara).
Erie	Buffalo, Black Rock, Dunkirk, Erie, Cleveland, Toledo, Monroe, and Detroit (on the Detroit river).
Huron	Port Huron, Bay City, Saginaw, Alpena, Cheboygan, Mackinaw.
Michigan	Traverse, Manistee, Père Marquette, Grand Haven, Chicago, Kenosha, Racine, Milwaukee, Sheboygan, Manitowoc, Green Bay.
Superior	Marquette, Houghton, Ontonagon, Superior City, Duluth.
Champlain	Rouse's Point, Plattsburg, Burlington.

Note 1.—A canal joins Champlain with the Hudson. Used chiefly by lumber and grain barges from Montreal, &c., to Albany.

2.—Calais and Eastport, on Passamaquoddy bay; Pembina and St. Vincent, on Red River; and Olympia, on Puget Sound, have considerable trade with Canada.

TWENTY-ONE CITIES WITH OVER 100,000 INHABITANTS.

Names.	Pop.	Names.	Pop.	Names.	Pop.
New York	1,207,000	Cincinnati	256,000	Louisville	124,000
Philadelphia	847,000	San Francisco	234,000	Cleveland	110,000
Brooklyn	587,000	New Orleans	216,000	Jersey	119,000
Chicago	504,000	Buffalo	156,000	Detroit	105,000
Boston	363,000	Washington	148,000	Milwaukee	104,000
St. Louis	351,000	Newark	137,000	Rochester	103,000
Baltimore	333,000	Pittsburg	125,000	Albany	102,000

Note 1.—New York, Maine, Massachusetts, Pennsylvania, own over two-thirds of the mercantile navy of the United States.

2.—The "Original Thirteen" are:—New Hampshire, Massachusetts, Rhode Island, Connecticut, New York, Pennsylvania, New Jersey, Delaware, Maryland, Virginia, North Carolina, South Carolina, Georgia.

3.—The "New England" States are:—Maine, New Hampshire, Vermont, Massachusetts, Rhode Island, Connecticut.

4.—New York, Pennsylvania, Ohio, Illinois, are the most populous states. They contain one-third of the entire population.

5.—The railway system of the United States extends to all parts of the country

Mexico.

Capital, Mexico (230,000). Population, 9,000,000, of whom less than one-third are of European origin.

BOUNDARIES.	*North* :—United States.
	East :—Gulf of Mexico, Campeachy Bay, and Caribbean Sea
	South :—Pacific Ocean, Guatemala, and British Honduras.
	West :—Pacific Ocean.
Rivers.	*East coast* :—Grande del Norte, Tampico.
	West coast :—Grande de Santiago (from lake Chapala), Sonora.
Cities.	*In the Interior* :—Mexico, Guadalaxara, Pueblo, Queretaro, Monterey.
	On the East coast :—Matamoras, Tampico, Vera Cruz, Merida, Campeachy.
	On the West coast :—Culiacan, Mazatlan, Acapulco, Tehuantepec.
Industries.	Mining, fruit-growing, cattle-raising, gathering dyes and medicinal-herbs, lumbering in mahogany and other cabinet woods, etc.

British Leewards

Capital, St. John, on Antigua.

Antigua, Montserrat, St. Christopher, Dominica, Barbuda, Nevis, Anguilla,
Virgins : Tortola, Virgin Gorda, Anegada.

British Windwards.

Capital, Bridgetown, on Barbadoes.

Barbadoes, St. Vincent, Grenada, Grenadines, Tobago. St. Lucia

---

French West Indies.

Guadaloupe, Martinique, Desirada, Marie Galante, St. Martin (in part), and Saintes.

St. Pierre, 20,000, and Port de France, or Port Royal, 13,000, are in Martinique.

Point-a-Pitre, 20,000, and Basseterre, are in Guadaloupe

Dutch West Indies.

Curacoa, Aruba, Bonaire, St. Eustache, Saba, and part of St. **Martin.**

EXAMINATION PAPERS.

Ontario Teachers' Examinations.

SECOND CLASS TEACHERS' AND INTERMEDIATE.

July, 1879.

1. Define equinox, steppes, great circle, and isothermal lines.

2. What is the form of the earth's orbit? How do you account for the **warmth of Summer in our** hemisphere, although the earth is farther from the sun than it is in Winter?

3. In what country, or countries, would one be most likely to find **the giraffe, the ostrich, the condor,** the reindeer, the chamois?

4. Outline the west coast of North America, indicating the islands near the coast, the rivers emptying into the Pacific ocean, and the principal cities on the seaboard.

5. Name the States bordering on the lakes between Canada and the United States, and mention at least two cities in each.

6. Name six rivers in Asia running south, also **the waters into** which they empty.

7. Over what railroads, and through what large towns or **cities, would** one pass **on a trip from Ottawa to Barrie?**

8. Where and what are Sitka, Cobequid, Lepanto, Cayenne, Socotra, Aral, Kertch, Wight, St. Louis, Canso, Tweed, and Cudleigh?

July, 1880.

1. Define estuary, river-basin, tropic, neap-tide, republic.

2. Explain the causes of Ocean Currents, and give the name and the course of *three* of the most important.

3. Trace **the Mississippi** River from its source to its mouth, naming the chief tributaries, East **and West, the States and chief towns bordering upon its banks,** and the principal commercial products for which it affords an outlet.

4. Sketch Europe from the Straits of **Dover to the Gulf of Genoa, indicating rivers, bays, capes, and** cities of importance along the coast.

5. Over what railroads, across what intersecting lines of railway, and through what cities and large towns, would you pass on a trip from Berlin to Amherstburg?

6. What and where are Ste. Maurice, Scugog, Rimouski, Chignecto, Pelée, Shediac, Burrard, Roanoke, **Galveston, and** the Cyclades?

7. Locate Cape St. **Lucas, Havana, Staten Island, Yapura River,** Jutland, Valparaiso, the Cambrian Hills, Cape Agulhas, Scilly Islands, **Table Bay, Warsaw, Baikal, Tonquin,** Ormuz, Loo Choo, and Zambezi.

July, 1880.

1. What are the natural divisions of South America? What the political?

2. State the principal causes which modify the climate of a country, and give examples.

3. How are **the frontiers between** Austria-Hungary **and** Turkey, and between Greece **and Turkey,** marked out?

4. Sketch the Atlantic coast line of the United States, marking the position of the chief capes and of the inlets, with the cities thereon.

5. Show how the latitude of a place is determined, and give the latitudes of New York, Toronto, Montreal, Florence, the Cape of Good Hope.

6. Describe (by a diagram if you can) the proposed route of the Canadian Pacific Railway, and show how it connects through Canadian territory with the Atlantic seaboard.

7. State the geographical position and the political relation of Candahar, Herat, Natal, Zanzibar, Hong. Kong, Corsica, Alsace.

8. State the form of government, and chief products, of Egypt, Brazil, Cuba, Bengal, Switzerland, and Cyprus.

9. What rivers flow from near the St. Gothard Pass in Switzerland, and what are their respective courses?

July, 1882

1. (i.) Name in order, beginning at the north and ending at Mexico, the Provinces of the Dominion and the States of the American Union on the eastern side of North America that possess one or more seaports; (ii.) name an important seaport in each ; (iii.) state the chief export or exports from each such seaport; and (iv.), if it is situated at the mouth of, or upon, a large river, name that river.

2. (i.) Contrast the physical characteristics of Northern and of Southern Europe.

(ii.) Arrange the governments of the different European States under the following heads : republics, limited monarchies, absolute monarchies.

3. (i.) Draw an outline map of Hindostan ; (ii.) mark on it the names and courses of three important rivers, and the names and positions of the chief mountain ranges and of four large cities.

4. (i.) Name five African lakes ; and (ii.) state with regard to each whether it is north of, south of, or on, the equator.

5. Explain why, though Canada is nearer the sun in January than in July, the weather is warmer in the latter month.

————

December, 1882.

1. In what countries are the sources and the mouths of the Elbe, the Meuse, the Douro, the Rhine, the Rhone, the Brahmaputra, the Yukon, the Columbia, the Colorado, the Amazon, the Vistula, and the Niemen?

2. State in detail what you would expect to see if you made a coast voyage around the Mediterranean

3. Draw a map of the British Isles, marking the courses of the Thames, the Severn, the Trent, the Tyne, the Tweed, the Clyde, the Shannon, and the Tay ; and the positions of Belfast, Dublin, York, Cork, Glasgow, Aberdeen, Dundee, Edinburgh, Liverpool, Manchester, Bristol, London, and Birmingham and of the smaller islands.

4. Compare the physical characteristics of Africa and of South America.

ENTRANCE TO HIGH SCHOOLS.

July, 1879.

1. Define crater, inlet, tropic, capital, and promontory.

2. Through what waters, and near what large cities, would you pass on a trip from Albany to Montreal, touching at Cape Race ?

3. Outline the coast of South America from Panama to Cape Horn, showing capes, rivers, &c., neatly printed in their proper places.

4. What and where are Scugog, Manitoulin, Hudson, Mobile, Pentland, Malar Medina, Lipari, Yapura, and Tchad ?

5. Suppose yourself at Winnipeg, with instructions to visit the capital of each Province lying eastward, describe your line of travel, naming railroads or water routes by which you would go.

6. Where, and how situated, are the following cities:—Kingston, Chicago, Boston, Halifax, New Orleans, Dublin, and St. Petersburg ?

————

December, 1879.

1. Define meridian, watershed, bay, frith, and zone.

2. What and where are Athabasca, Nelson, Chignecto, Restigouche, Gatineau, Temiscaming, St. Hyacinthe, Quinte, Chesapeake, Sacramento, Champlain, and New Orleans?

3. Where do you find the following natural productions in great abundance :—Cotton, copper, coal, coffee, tin, gold, furs, and grapes ?

4. Say you embark at the Isle of Man on a voyage to the mouth of the Volga. Through what waters, and near what capes and islands, would you pass ?

5. Draw a map of the coast of Asia, from Behring Straits to Cape Comorin, showing all the important physical features with their names neatly printed upon them.

6. Locate the following :—Obi, Papua, Zambezi, Tunis, Morea, Cyprus, Venice, Lyons, Copenhagen, Borneo, Cheviot Hills, Crimea, Quito, Port-au-Prince, Trinidad, and Loffoden.

July, 1880.

1. Define watershed, frith, delta, horizon, axis of the earth, polar circles, ecliptic, first meridian.

2. (a) Why are the days in the northern hemisphere longer in summer than in winter?

(b) What causes the change of seasons?

(c) Why does the sun appear to set in the west?

3. Trace the following rivers from sources to outlets, and name the principal cities on their banks:—Danube, Rhine, Ganges, St. Lawrence, Mississippi.

4. Name the cities of Ontario, and give the situation of each.

5. Over what railroads would you pass in going (I.) from Hamilton to Peterboro; (II.) from Collingwood to London?

6. What are the chief natural products of Manitoba, Nova Scotia, Southern States of America, France, China?

7. Where are the following:—Islands—Malta, Anticosti, Ceylon? Capes—Verde, Comorin, La Hogue? Bays—Verte, All Saints, Table?

———

December, 1880.

1. Define isthmus, promontory, beach, bay, inlet, sound, roadstead, strait.

2. Name and give the boundaries of the zones. What determines the two tropics and the two polar circles?

3. Define latitude, longitude, first meridian. What is the greatest latitude a place can have? The greatest longitude? Why?

4. Give, with their boundaries, the political divisions of North America.

5. Name, giving their relative positions, the divisions of British North America. Which of these are comprised in the Dominion of Canada, and what are their capitals?

6. Make a list of the principal rivers of Ontario, telling into what bodies of water each flows.

7. Give the boundaries of Asia, and the relative positions of its chief political divisions.

8. Draw an outline map of Ireland, and mark the positions of Dublin, Belfast, Cork, Limerick.

———

July, 1881.

1. Define physical geography, plateau, river-basin, watershed, meridian, zone; absolute monarchy, republic.

2. Name the Provinces of Canada, giving their relative positions. Also, give the name and position of the capital of each Province.

3. Of what lakes are the following rivers the outlets:—Nelson, Detroit, Severn, Richelieu, Saguenay, San Juan, Rhine, Rhone?

4. Name, in order, the seas, gulfs, bays, and straits of Europe.

5. Give, as definitely as you can, the position of the following cities:—Chicago, Buffalo, St. Catharines, St. John, Rio Janeiro, Hull, Manchester, Glasgow. Islands—Skye, Funen, St. Helena, Cyprus. Mountains—Blanc, Cotopaxi, Vesuvius, St. Elias.

6. What are the chief productions of France, Barbary States, Hindostan, Nova Scotia, Gulf States of North America, Central America?

7. A vessel carries freights between Montreal and Cuba. What will her cargo probably be (1) on her outward trip; (2) on her return trip?

8. By what railroads would you travel in going:—

(1) From Hamilton to Peterboro'?

(2) From Ottawa to Barrie?

———

December, 1881.

1. What is political geography? physical geography? Define the following:—First meridian, zone, equinox, plateau, watershed, glacier, climate.

2. Give the boundaries of the different zones, and the breadth of each zone in degrees. Account for the positions of the bounding lines of the zones.

3. What and where are the following:—Vancouver, Three Rivers, Trinidad, Avon, Corfu, Mersey, Cromboll, Hamburg, Hindoo Koosh, Lyons?

4. Name the bodies of water into which the following rivers flow :—Garonne, Tagus, Elbe, Volga, Oder.

5. Between what cities in the United States and in the British Islands is trade with Canada chiefly carried on ? Tell what you know of the commodities exchanged.

6. Over what railroads would you pass in going from (1) Toronto to St. Thomas ; (2) Owen Sound to Ottawa ? Describe a trip from Montreal to Lake Superior.

7. What information respecting a country can be obtained from a knowledge of its mountains ?

8. Name and classify according to slopes the principal rivers of Asia.

9. From what countries do we chiefly obtain the following :—Coal, iron, cotton, rice, sugar, coffee, silk, opium ?

June, 1882.

1. Give the boundaries of South America, and name its extreme northern, eastern, southern, and western points.

2. What is the most important mountain system in Central Europe ? Name four rivers rising in these mountains, and the body of water into which each flows.

3. State the position of the following :—

　Cities— Edinburgh, Liverpool, Paris, Vienna, Calcutta.

　Islands—Corsica, St. Helena, Sardinia, Cyprus, Madagascar.

　Capes—Race, Cod, Matapan, Guardafui, Comorin, Land's End, Sable, Clear.

　Gulfs, Bays, and Straits—Fundy, Panama, Lyons, Baffin, Biscay, Venice, Messina, Bonifacio.

4. Name the cities of Ontario, giving as nearly as you can the population of each.

5. Describe the courses of the chief rivers of Ontario. Give the position of Windsor, Sarnia, Kincardine, Collingwood, Toronto, Kingston.

6. Draw an outline map of North America, marking the position of Boston, New York City, Halifax, St. John, Montreal, San Francisco, and New Orleans.

Nova Scotia Teachers' Examinations.

B　1879.

1. What do you understand by climate ? What effect on it have (1) cultivation of the soil and drainage ; (2) the clearing of forests ? Give reasons for answer to latter.

2. State generally received theory with respect to the formation of dew. Has it been called in question ?

3. Define equator, ecliptic, equinoxes, solstices, circle of illumination, declination.

4. How do you find the length of a degree of longitude in any given latitude ? Explain reasons of rule. (Globe.)

5. Name the foreign possessions of France, in the order of their importance.

6. Give the boundaries of Afghanistan, with the approximate latitude and longitude of the extreme southern and western points.

7. Locate as definitely as possible the following places :—Toledo, Heidelberg, Omaha, Aden, Auckland, Duluth, Calicut.

8. Describe the German Empire, (1) as to organization and government ; (2) as to soil, climate, and products.

1. Name the chief canals of the Dominion, stating the waters connected by them, and the reason of their construction.

2. Name in order of population the nine largest cities of the Dominion of Canada.

3. Name the chief ports of the Dominion from which lumber is exported.

4. Name in order (1) the six longest rivers in British America ; (2) the four chief rivers of British America, in order of commercial importance.

5. Draw an outline map of the Maritime Provinces of the Dominion, marking the chief rivers, and locating the county towns of Nova Scotia.

B. 1880.

1. Define watershed, river-system, mountain-chain; and account for the names Tropic of Cancer, Tropic of Capricorn, and Arctic Circle.

2. Describe the Austro-Hungarian Empire, (1) as to government and races; (2) as to physical forces and resources.

3. Where, and for what noted, are Trincomalee, Bushire, Silistria, Cordova, Timbuctoo, Pittsburg, Cusco, Paramatta, Brindisi?

4. Describe the tides, their cause, the extent of the movement, and the tidal periods.

5. How do you find the duration of twilight at a given place on a given day? (Globe.)

6. Mention some noteworthy features of the general contour of the continents.

7. Name the chief rivers of the United States, with the principal cities on their banks.

---

1. Name the coast waters, capes, and islands of Newfoundland.

2. Describe the North-West Territories, (1) as to physical features; (2) as to natural resources.

3. Name the counties of Quebec north of the River St. Lawrence.

4. Write brief notes on the leading industries of the Maritime Provinces.

5. Draw an outline map of the Province of Ontario, locating the chief cities, and marking the adjacent waters.

Manitoba Teachers' Examinations.

1879.

1. Define the ecliptic and polar circles.

2. Explain the causes of solar and lunar eclipses.

3. Give the boundaries and principal divisions of North America. Name also its principal islands.

4. Trace the courses of the Rocky and Alleghany Mountains, stating where they each attain their highest summits.

5. Name the chief divisions of South America, and its most important rivers, giving their length and sources.

6. Draw a sketch of the coast of Europe, giving the extent of its coast line.

7. Give the extent and population of Germany. Describe also its government.

8. Describe the situation of Afghanistan, giving its capital and chief passes.

9. What islands are comprised in Australasia? Describe the largest of them.

1880.

1. Explain fully the motions of the earth. Define peninsula, cape, watershed, ocean, sea.

2. Give the principal islands and capes of North America.

3. Name the lakes and rivers of the Province of Ontario.

4. How is the United States of America bounded? Describe the mountains of the United States.

5. Give the seas and gulfs of Europe.

6. Describe the position of Nova Scotia. Give its counties.

7. Mention the principal towns of Spain. Describe the position of Gibraltar.

8. What does Asia Minor comprise? Give the provinces of British India with their capitals.

9. Give the boundaries of Africa. Name its countries.

10. How is Australia divided? Describe one of its divisions.

1880.

　1. In what countries are tin, mercury, and lead found? Define latitude and longitude, and explain their use.

　2. Name the chief influences which affect the climate of a country.

　3. Give the boundaries of the Dominion of Canada, and name the principal rivers of the Dominion, mentioning the province or territory in which they are situated.

　4. Name the countries of South America and describe its mountains.

　5. Describe the course of the Danube and state what cities are situated on it.

　6. What mountains separate Switzerland from France? Give the islands of Europe.

　7. Give the principal capes and gulfs of Asia.

　8. Describe New South Wales. Name its chief towns.

　9. Describe the position of the following places :— Benares, Prague, **Lyons, Yokohama, Belfast, Cairo, Dantzic,** Dunedin, Oporto.

　10. Give the British, French, and Portuguese islands of Africa. Describe Abyssinia fully.

1881.

　1. Draw a map of South America, defining the areas drained by the **principal rivers, and stating the** positions of the chief cities.

　2. Name the islands of the West Indies and describe them physically, politically, and commercially.

　3. Trace the principal watersheds of North America, and mention the groups into which its rivers may be divided.

　4. What are the principal islands in the Pacific and what is their importance?

　5. Name the rivers emptying into the Indian ocean and describe the course of two **of them.**

　6. Name the British possessions in Asia.

　7. State the relative positions of the British possessions in Africa.

　8. Name the American States east of the Mississippi, with the capital of each.

　9. Draw an outline map of the Dominion of Canada.

　10. Give a list of the exports of Canada and of the countries to which they are sent.

　11. Name the Cities of the Dominion and give the population **of each.**

　12. How do you find the distance between two places on **an ordinary globe?**

　13. How do you find the latitude of a place **on the globe?**

　14. What are the lines usually found on globes?

　15. Account for seasons, winds, tides, and ocean currents, **and trace the course of** movement of the atmosphere and the **ocean.**

---

　1. Explain perihelion, apogee, solstice, zodiac, ecliptic.

　2. Classify tides. State their causes. Draw diagrams to illustrate.

　3. Show clearly why the Arctic Circle is 23½° (nearly) from the Pole.

　4. If the earth's axis should make an angle of 30° with a perpendicular **to the plane of its orbit, what** changes would follow?

　5. What is the jurisdiction of the Dominion Government as distinguished from **that of the Provincial** Governments?

　6. Where does Canada get supplies of carpets, sugars, lace, cork, canned salmon, and wine?

　7. Name the Mediterranean seaports of Egypt, of Spain, and of France.

　8. Name the cities on the Rhine and on the Danube.

　9. Name the foreign possessions of France, of Portugal, and of Holland.

　10. What are the form of government and the capital of each of the following countries :—Egypt, Chili, Australia, Russia, England?

　11. What and where are the following :—Gallinas, Pamlico, **Socotra, Elsinore, Calgary,** Prague, Riga, Miquelon, Auckland, Yellow Head?

　12. To what nation do the following belong :—New Guinea, Hayti, Society Islands, Celebes, Heligoland, Sumatra, Ushant, **Puerto Rico, Jersey** Islands, Philippine Islands?

　13. Name the lake expansions of the Shannon and of the Ottawa rivers.

British Columbia.

Area, 341,000 square miles ; greatest extent from north to south, 750 miles, and from east to west, 500 miles.

Vancouver Island.—Greatest length, 275 miles ; greatest breadth, 90 miles ; average breadth, 60 miles.

Population—150,000 (approximate), of whom about 30,000 are Indians.

BOUNDARIES.

*North:—*Northwest Territories, or the parallel of 60° N. latitude.

*East:—*Northwest Territories, or the Rocky Mountains and the line of 120° W. longitude.

*South:—*United States, or the parallel of 49° N. latitude.

*West:—*Pacific Ocean, and a strip of coast line forming part of Alaska.

Government.

1. *Legislative:—***Legislative Assembly**, consisting of :—
   - (a) Thirty-three members *elected* by the people.
   - (b) A Lieutenant-Governor *appointed* by the Privy Council of Canada.

2. *Executive:—*An **Executive Council** composed of the Lieutenant-Governor and leaders of the ruling party in the Assembly. These must be members of the Assembly.

   The *revenue* is derived chiefly from the following sources:—Dominion of Canada, the sale of Crown grants and timber limits, licenses, taxes, &c.

   The *expenditure* is chiefly on account of :—Civil Government, education, administration of Justice, public works, &c.

Lakes.

*On the mainland:—*Okanagan, **Kootenay**, Osoyoos, Shuswap, Quesnelle, François, Stuart, Babine.

*On Vancouver Island:—*Sooke, **Cowichan**, **Quamichan**, **Shawnigan**, **Great Central**, Ninnkish.

Gulfs or Straits. Inlets.

Georgia, Juan de Fuca, Haro, Queen Charlotte Sound.

*The mainland—*Burrard, Bute, Howe Sound, **Jervis**, Knight, Rivers, Bentinck, Arm, Milbank Sound, Gardiner, Douglas Channel, Observatory.

*Vancouver Island—Sounds:—*Barclay, Nootka, Quatsino, Clayoquot, Baynes.

Rivers.

*Into the Pacific:—*Columbia, Bella Coola, Salmon, Skeena, Naas, Stickeen.

*Into the Gulf of Georgia:—*Fraser (1,000 miles).

*Into the Fraser:—*Thompson, Bridge, Chilcotin, Quesnelle, Stuart.

Peace, into Lake Athabasca.

Kootenay, into Columbia.

Finlay,   "   Peace.

Liard,   "   Mackenzie.

Islands.

Vancouver, Queen Charlotte Group, Salt Spring, Gabriola, Denman, Mayne.

Mountains.

1. **Rocky**—*Highest Peaks:—*Brown (16,500 feet), Hooker, Murchison.

   *Chief Passes are:—*Kicking Horse, Yellowhead, Vermillion, **Kootenay**, Peace River.

2. *Cascade:—*In the western part of the mainland ; no very high peaks ; terminates in Mt. St. Elias on boundary line of Alaska.

3. *Selkirk—Gold:—*Ranges parallel with the Rocky Mountains.

4. *Coast Range:—*Extending through the islands **on the** west coast, has for its highest peak Mt. Victoria (about 8,300 feet), on Vancouver Island.

   These mountains inclose the valleys of the **Fraser**, Thompson, Columbia, Skeena, Stickeen, and Peace Rivers.

Surface.	Generally mountainous. **The mountains are covered with** forests of Douglas pine, fir, balsam, hemlock, cedar, and other woods. The Douglas pine, which often **grows** to the height of 300 feet, is largely exported, being especially valuable for masts of ships, &c.
	In the interior of the mainland are vast tracts of land which afford every facility for stock-raising. These lands, if irrigated, would become excellent agricultural districts.
	The soil is very fertile, producing cereals, roots, &c. similar to those of the other provinces, and is especially adapted to fruit culture.
Climate.	Owing to the warm Japan current, it is milder in the islands than in any other part of the Dominion; on the mainland it is dry and subject to extremes of heat and cold.
Industries.	Mining, lumbering, fishing, stock-raising, agriculture.
Minerals.	The principal gold-fields are in Cariboo, Kootenay, and Cassiar. Silver, copper, iron, &c. abound.
	Coal is found in all parts of the province, but the principal mines are located in Nanaimo and Wellington; the yield, being of excellent quality, is largely exported.
Cities, Towns &c.	*Vancouver Island:*—Victoria (12,000), **capital**; **Nanaimo, Esquimalt,** Wellington, Comox, Alberni.
	The mainland:—New Westminster, Vancouver, Port Moody, Yale, Lytton, Ashcroft, Kamloops, Revelstoke, Donald, Clinton, Lillooet, Quesnellemouth, Barkerville.
Railways.	The Canadian Pacific; branch to New Westminster. The Esquimalt and Nanaimo; from Victoria to Nanaimo.
Naval Station	Esquimalt, possessing a splendid harbor (whose entrance is Royal Roads), and containing a fine dock-yard, is Her Majesty's naval station for the North Pacific.
Indians.	Chiefly located on reservations. They gain a livelihood by fishing, hunting, trapping, and agriculture.
	Principal tribes are:—Flatheads, Fort Ruperts, Bella Bellas, Bella Coolas, Tsimpsheans, Hydahs, Shuswaps, Kootenays, Carriers, Chilcotins.
Divisions.	The province is divided into sixteen electoral districts, as follows:—

Vancouver Island	The Mainland
Victoria City.	New Westminster City.
Victoria.	New Westminster.
Esquimalt.	Yale.
Cowichan.	Lillooet.
Nanaimo.	Cariboo.
Comox.	Kootenay.
	Cassiar.

On the discovery of extensive gold fields in 1858, in this part of Canada, two Crown colonies were formed, viz.: Vancouver Island and British Columbia. These colonies were united in 1866. In 1871 British Columbia became a province of the Dominion.

Although the least of the provinces in population, it is the greatest in extent, and is the first in exports in proportion to population, as well as in the amount expended for education compared with revenue.

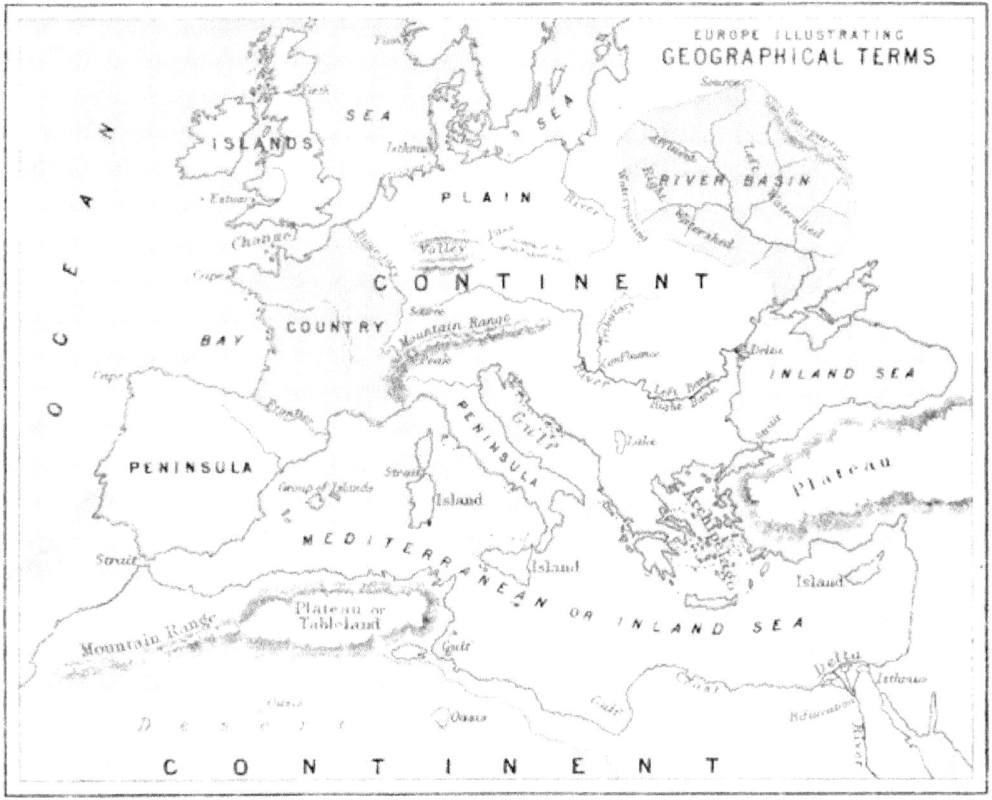

EUROPE ILLUSTRATING
GEOGRAPHICAL TERMS

RIVER BASIN

INLAND SEA

PLATEAU

Archipelago

CONTINENT

SEA

SEA

PLAIN

GULF

PENINSULA

MEDITERRANEAN OR INLAND SEA

CONTINENT

ISLANDS

BAY

COUNTRY

Mountain Range

Island

Island

Island

PENINSULA

Mountain Range

Plateau or
Table-land

DESERT

CONTINENT

OCEAN

River

River

River

Valley

W. J. Gage & Co Toronto

2

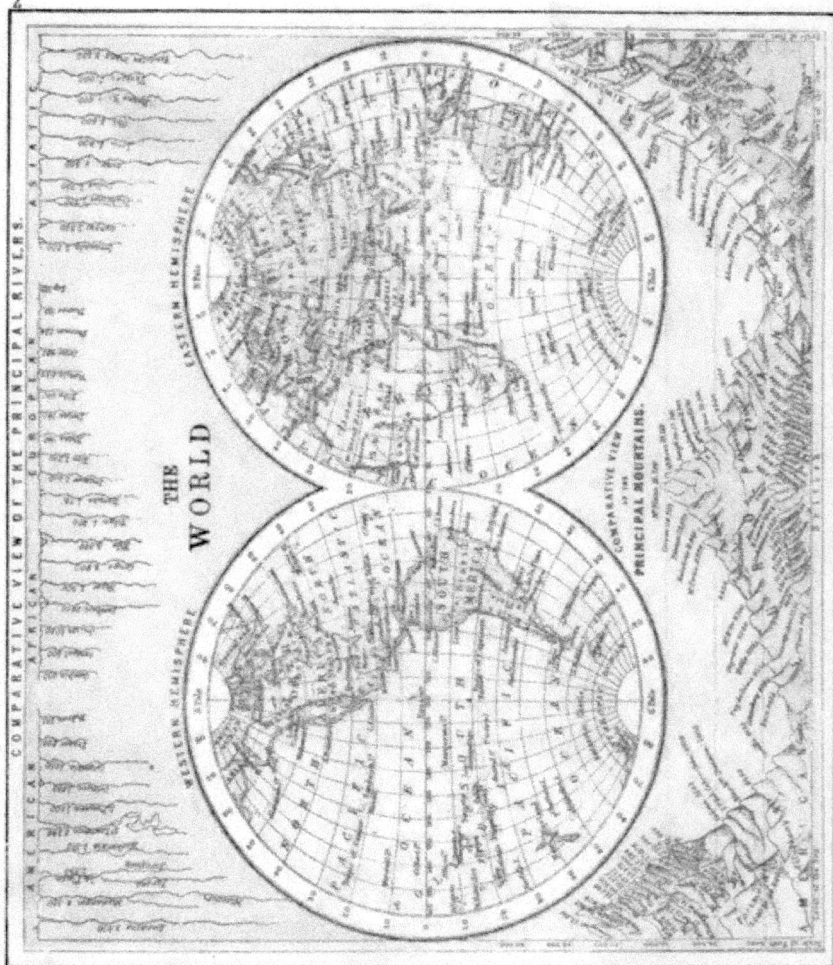

THE WORLD

EASTERN HEMISPHERE

WESTERN HEMISPHERE

COMPARATIVE VIEW OF THE PRINCIPAL RIVERS.

AMERICAN   AFRICAN   EUROPEAN   ASIATIC

COMPARATIVE VIEW OF THE PRINCIPAL MOUNTAINS.

W. J. Gage & Co., Toronto.

SOUTH AMERICA

English Miles

BRITISH STATES OF
COLOMBIA

ECUADOR

PERU

BRAZIL

BOLIVIA

ARGENTINE
REPUBLIC

URUGUAY
or
BANDA ORIENTAL

BUENOS AYRES

Rio de la Plata

ATLANTIC OCEAN

PACIFIC OCEAN

Equator

Falkland I.

South
Georgia I.

W. J. Gage & Co. Toronto.

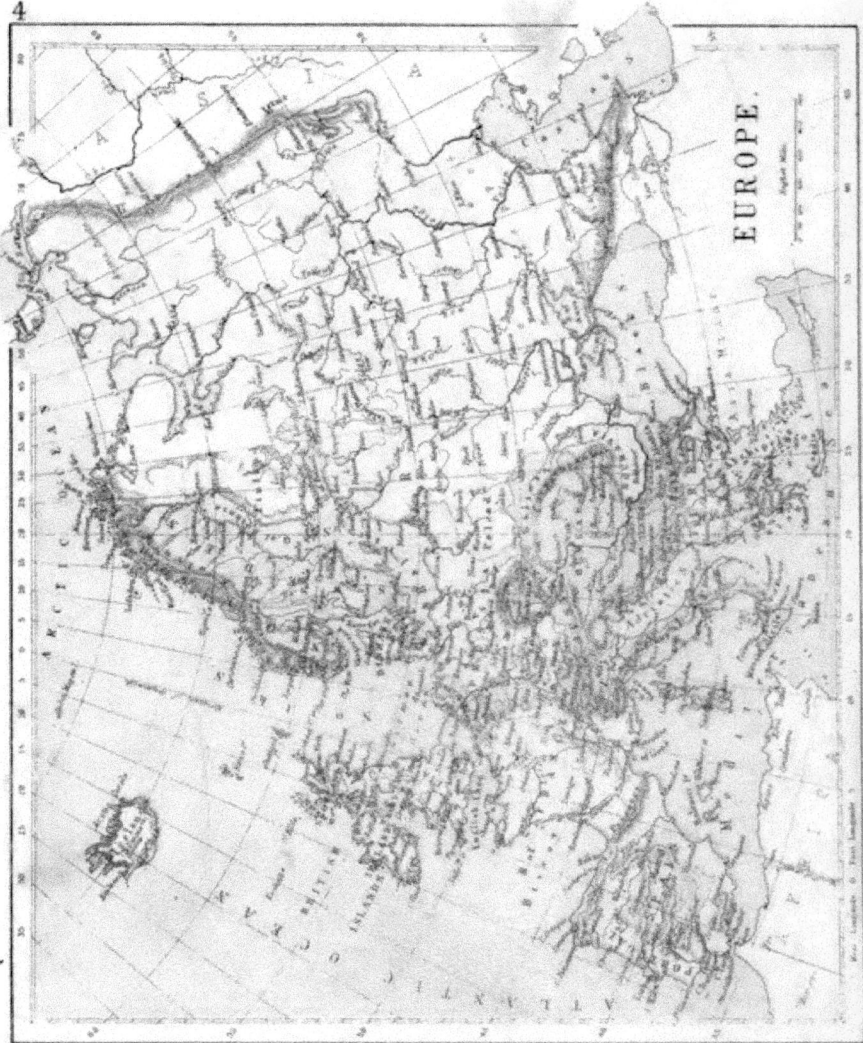

EUROPE.

ASIA

ARCTIC OCEAN

ATLANTIC OCEAN

AFRICA

ENGLAND & WALES

English Miles

W. J. Gage & Co. Toronto.

SCOTLAND

DOMINION OF
CANADA

ATLANTIC OCEAN

ARCTIC OCEAN

HUDSON BAY

PACIFIC OCEAN

UNITED STATES

W. J. Gage & Co., Toronto

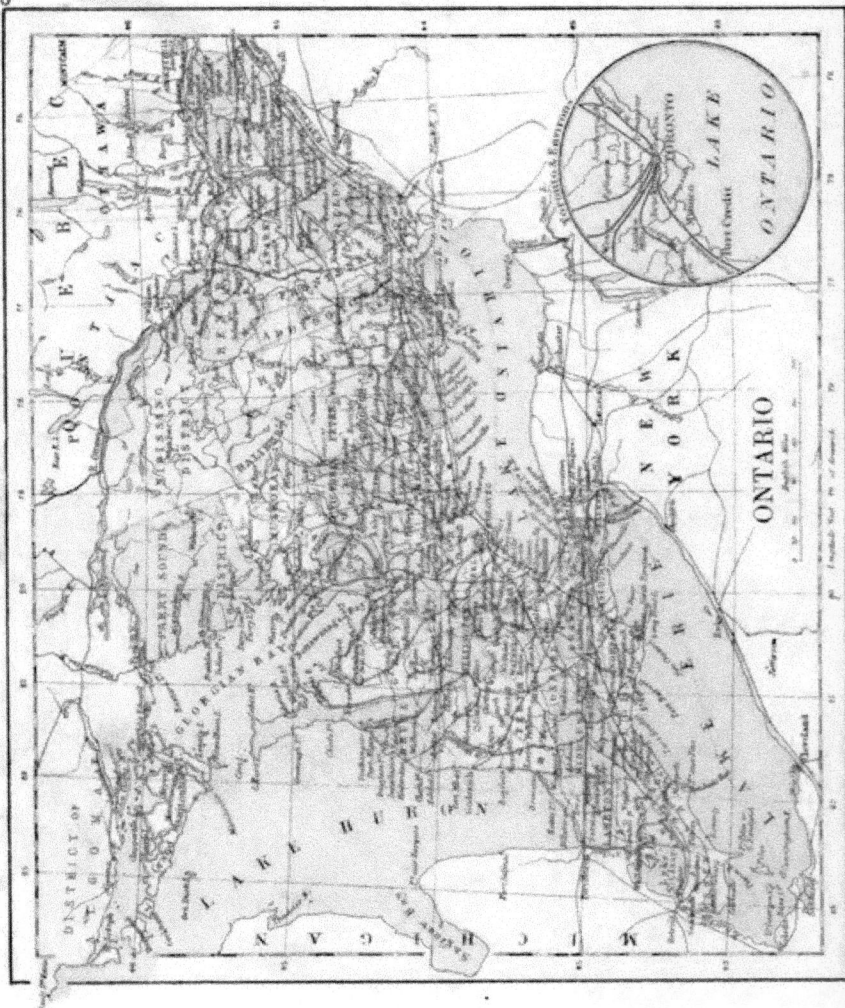

ONTARIO

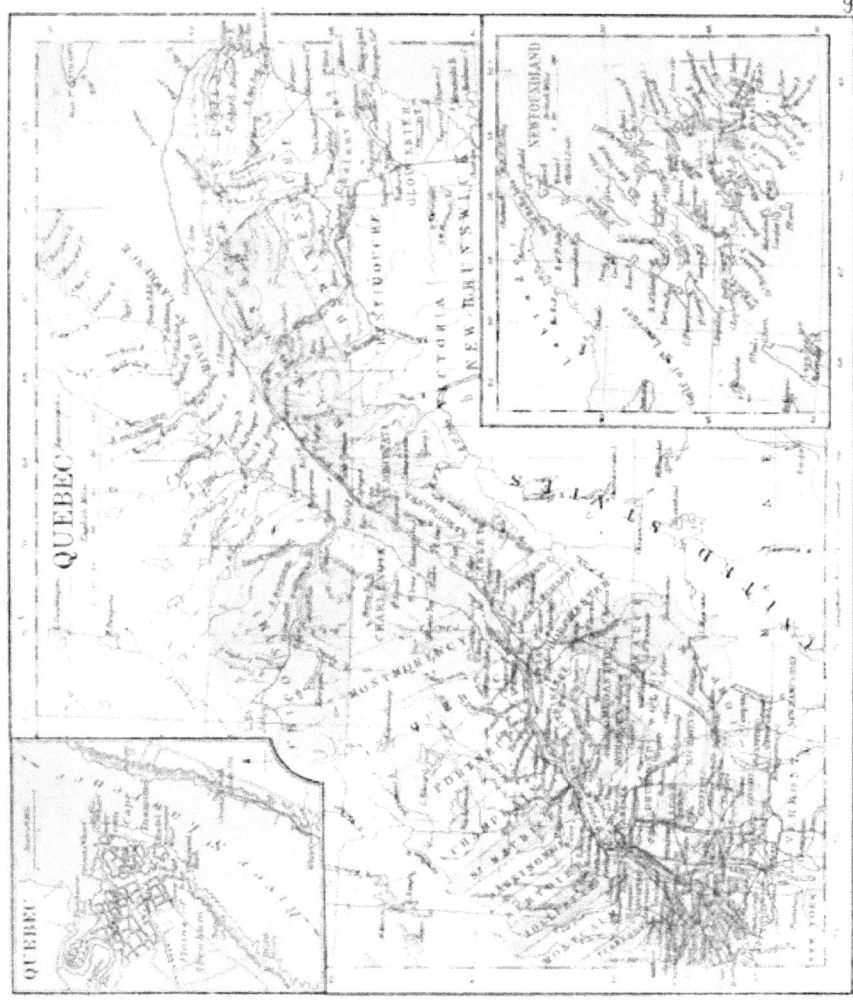

QUEBEC

NEWFOUNDLAND

NEW BRUNSWICK

QUEBEC

W.J. Gage & Co. Toronto.

9

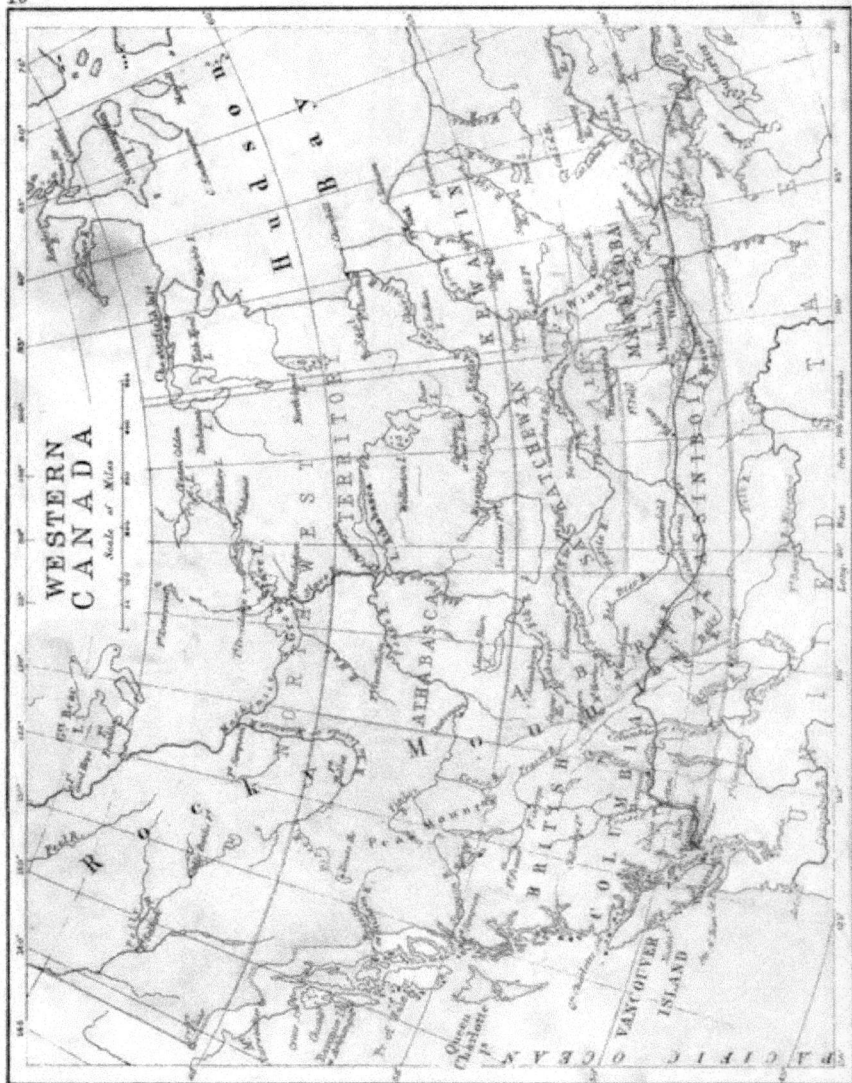

WESTERN
CANADA

Scale of Miles

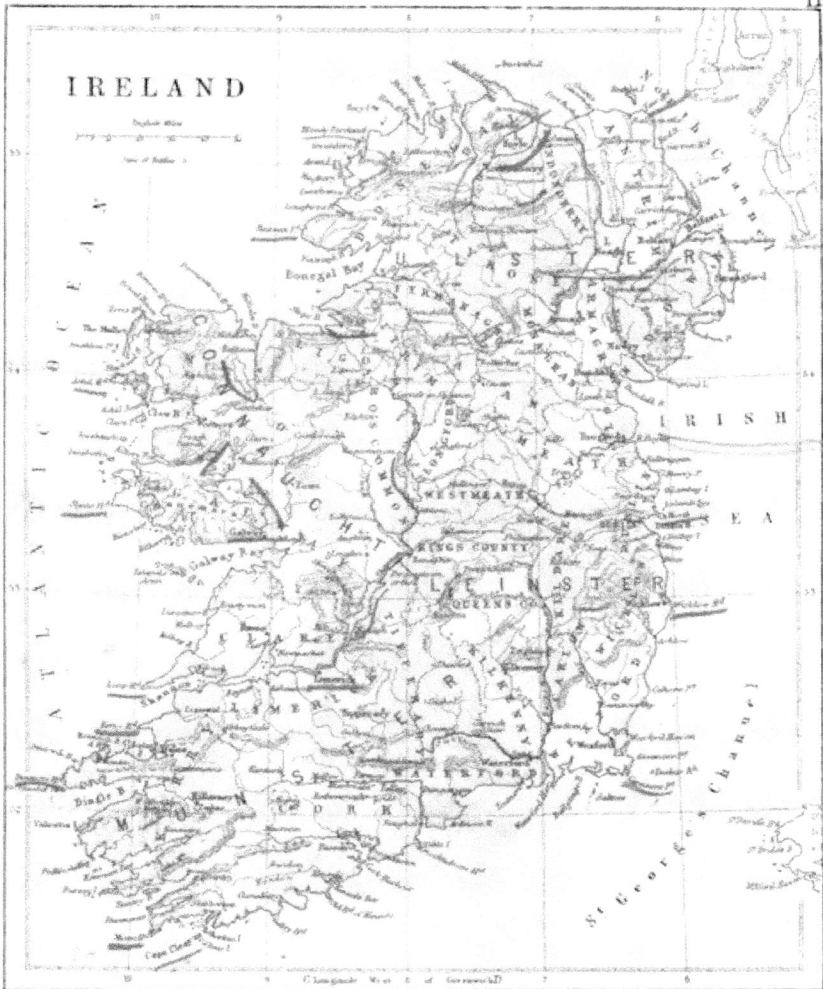

IRELAND

ATLANTIC OCEAN

Donegal Bay

IRISH

SEA

CONNAUGHT

WESTMEATH

KINGS COUNTY

LEINSTER

QUEENS COUNTY

KILKENNY

LIMERICK

MUNSTER

CORK

Dingle B.

St George's Channel

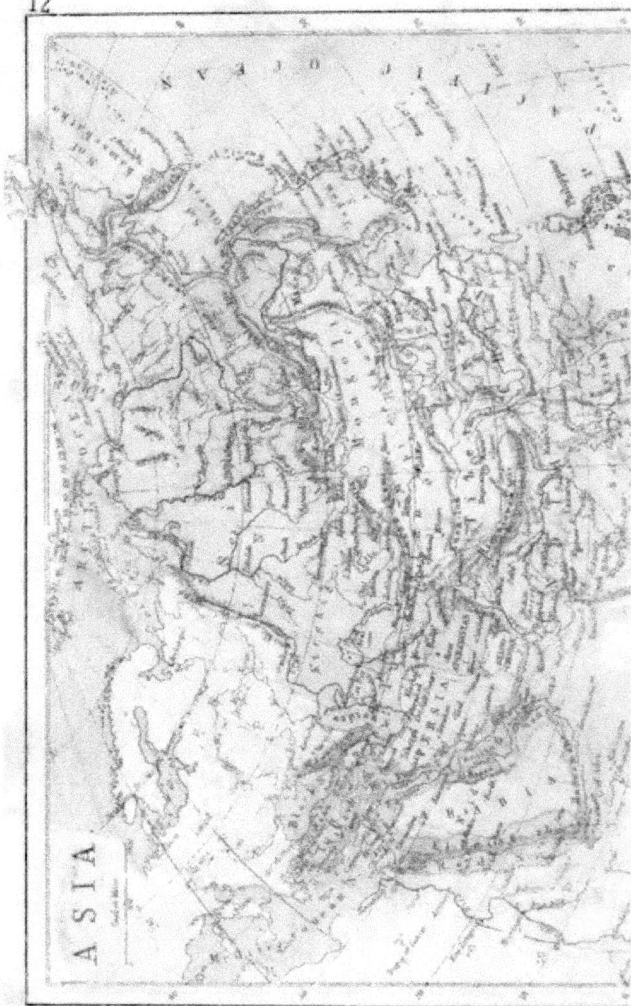

ASIA

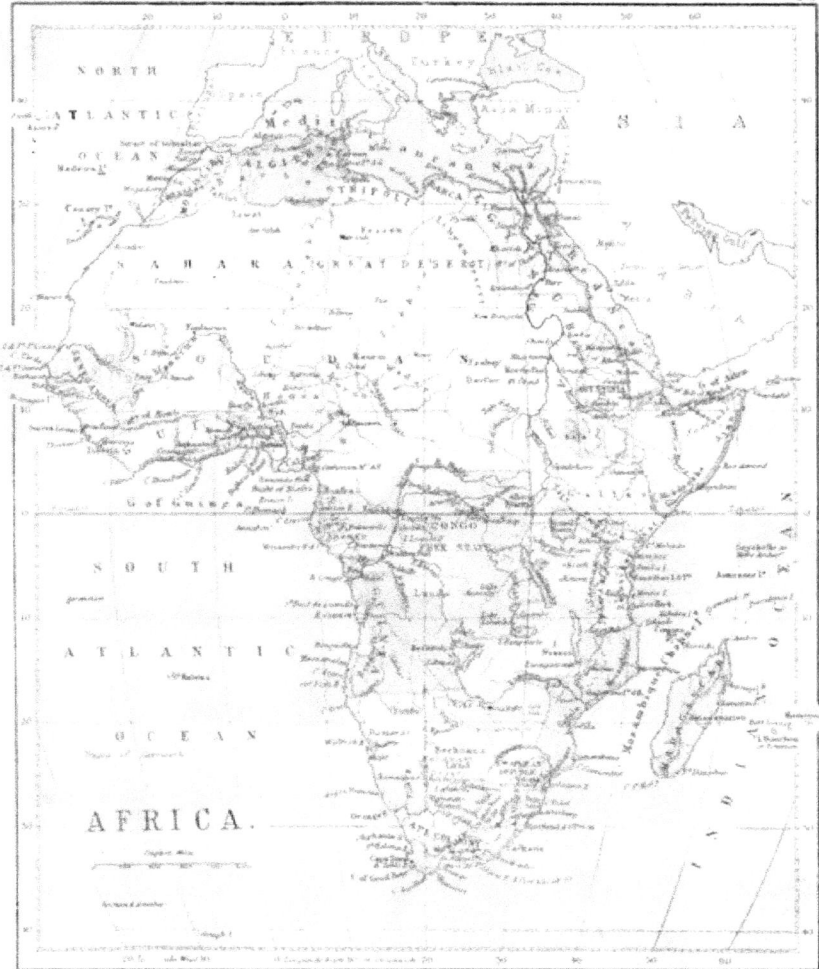

AFRICA.

W. J. Gage & Co., Toronto

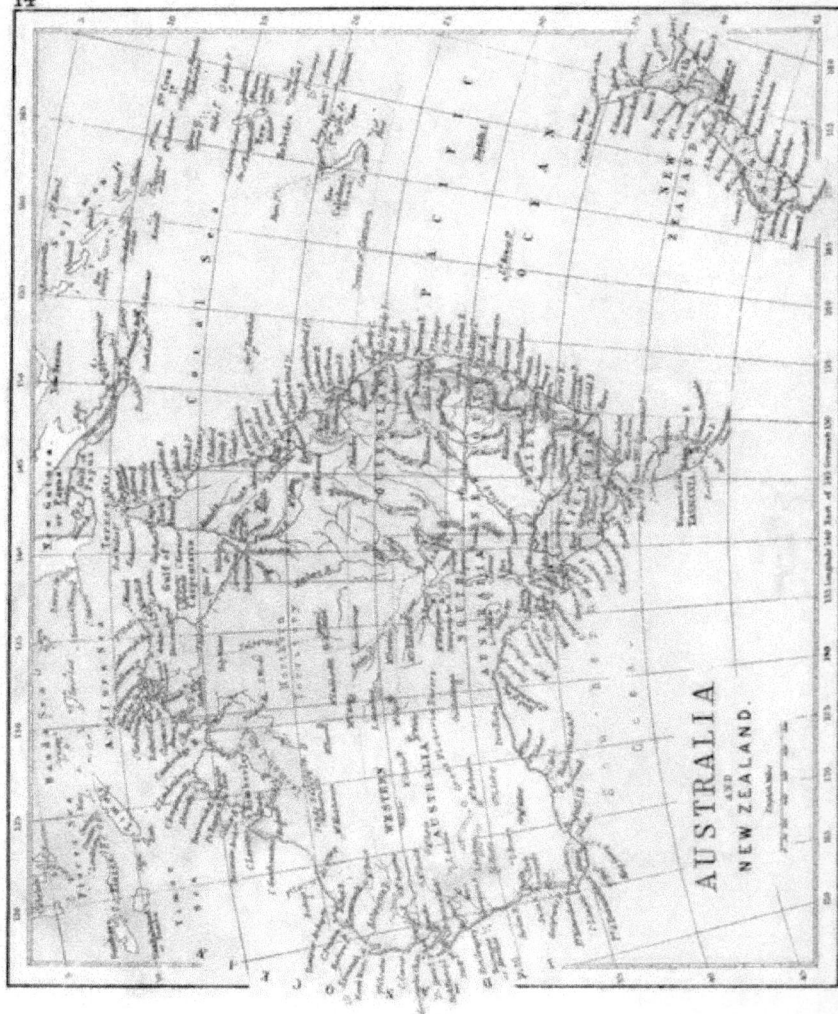

AUSTRALIA
AND
NEW ZEALAND.

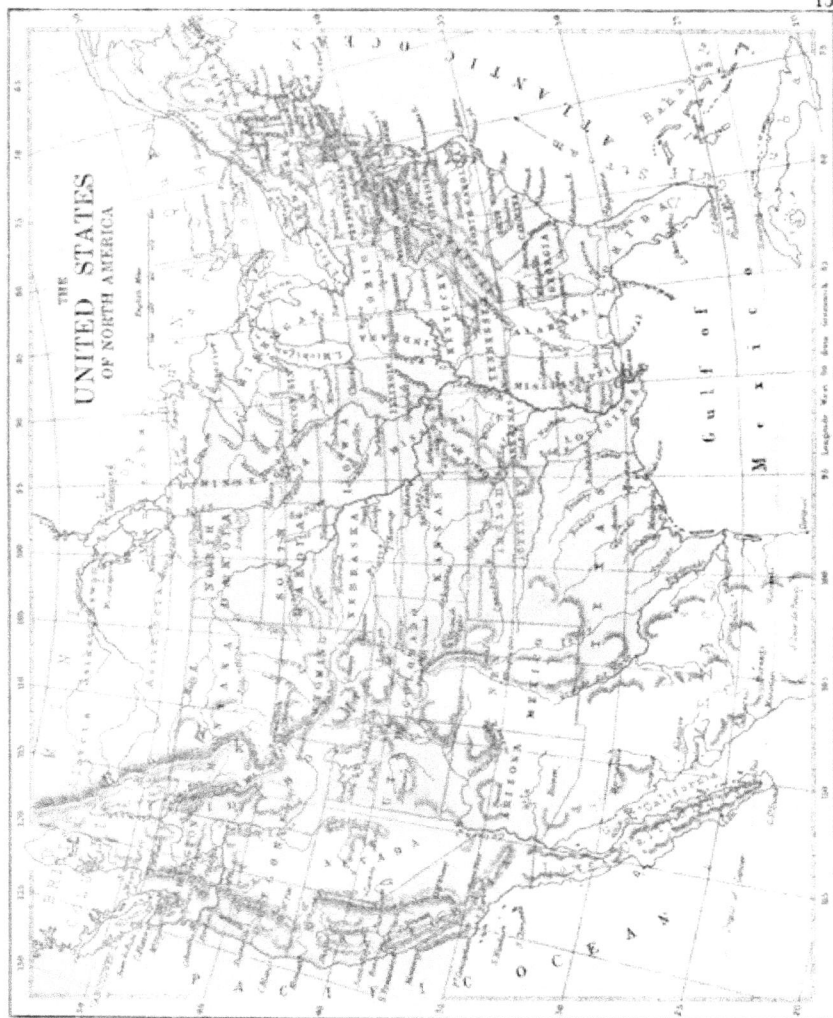

THE
UNITED STATES
OF NORTH AMERICA

W. J. Gage & Co. Toronto

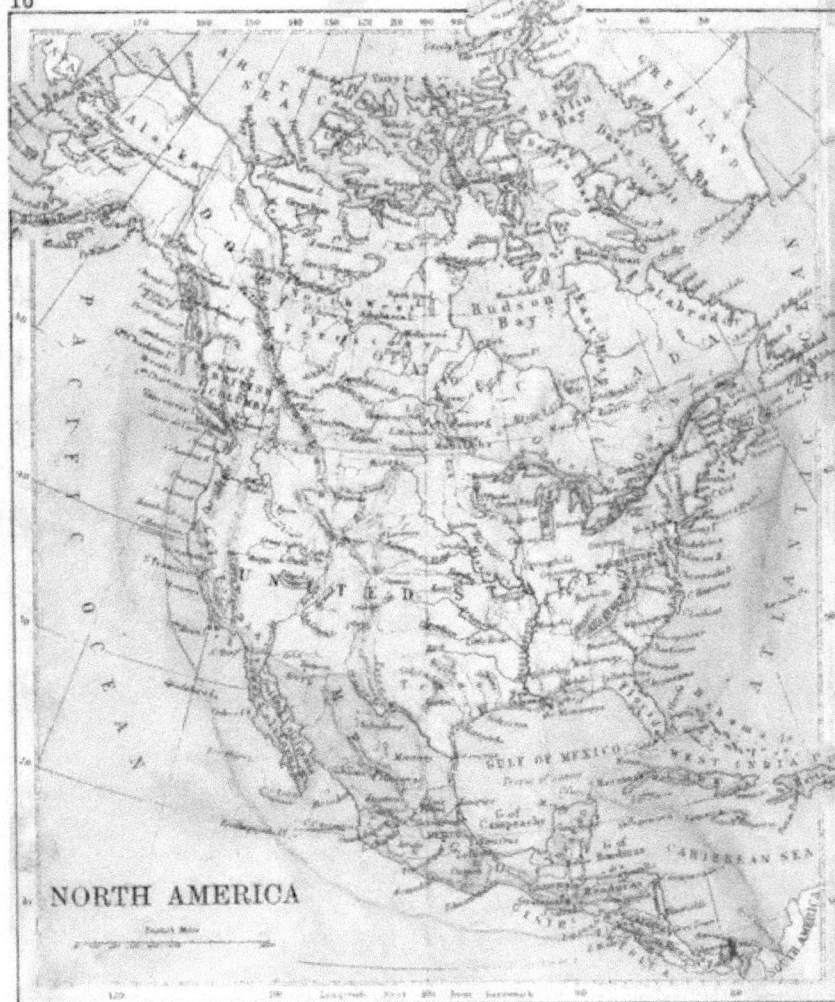

NORTH AMERICA

W. J. Gage & Co., Toronto.